AF361916

MEMS Sensors and Resonators

MEMS Sensors and Resonators

Special Issue Editor
Frederic Nabki

MDPI • Basel • Beijing • Wuhan • Barcelona • Belgrade • Manchester • Tokyo • Cluj • Tianjin

Special Issue Editor
Frederic Nabki
École de Technologie Supérieure
Canada

Editorial Office
MDPI
St. Alban-Anlage 66
4052 Basel, Switzerland

This is a reprint of articles from the Special Issue published online in the open access journal *Sensors* (ISSN 1424-8220) (available at: https://www.mdpi.com/journal/sensors/special_issues/MEMS_SENSORS).

For citation purposes, cite each article independently as indicated on the article page online and as indicated below:

LastName, A.A.; LastName, B.B.; LastName, C.C. Article Title. *Journal Name* **Year**, *Article Number*, Page Range.

ISBN 978-3-03928-865-6 (Pbk)
ISBN 978-3-03928-866-3 (PDF)

Cover image courtesy of Frederic Nabki.

Contents

About the Special Issue Editor

Frederic Nabki, Prof. Dr., received a B.Eng. degree (Hons.) in electrical engineering and a PhD in electrical engineering from McGill University, Montreal, QC, Canada, in 2003 and 2010, respectively. He is a Full Professor of Electrical Engineering at the École de technologie supérieure (ETS) in Montreal, Canada. His research interests include microelectromechanical systems (MEMS) and analog, RF, and mixed-signal integrated circuits, specifically focusing on the creation of next generation MEMS processes and devices, the integration of MEMS devices with CMOS systems, the modeling of MEMS devices, and the design of integrated circuits such as analog to digital converters, sensor interfaces, and ultra-wideband wireless transceivers. He has co-authored two book chapters and over one hundred publications, as well as holding 11 issued patents and 21 pending patent applications related to MEMS and CMOS/MEMS monolithic integration. His research interests include MEMS and RF/analog microelectronics. He was a recipient of the Governor General of Canada's Academic Bronze Medal, the J.J. Archambault IEEE Canada Medal, and the UQAM Faculty of Science Early Career Research Award. He is supported by the Microsystems Strategic Alliance of Quebec (ReSMiQ), the Quebec Fund for Research in Nature and Technology (FRQNT), the Ministry of Economy, Science and Innovation (MESI) of Quebec, the Natural Sciences and Engineering Research Council of Canada (NSERC), and the Canada Foundation for Innovation (CFI).

Preface to "MEMS Sensors and Resonators"

MEMS Sensors and Resonators: Infinite Possibilities of the Infinitely Small

Like integrated circuits (ICs), microelectromechanical systems (MEMS) are a disruptive technology enhancing systems and enabling new applications. The MEMS market is expanding at an increasing rate and represented, in 2018, an US\$1.6 billion market globally and projected to grow to US\$18 billion by 2024 [Yole, 2019]. Recently, MEMS sensors have been increasingly present in wearables (e.g., smartwatches and fitness trackers) and sensor nodes (e.g., tracking and weather forecasting), where many sensors are combined. This warrants tighter integration to allow for more compact sensing systems. Such sensors are suited to the Internet of Things, where large numbers of small and autonomous sensor nodes, independent of power and wiring infrastructure, are expected to be deployed with much-reduced overhead.

This Special Issue touches on some of the global works on MEMS sensors and resonators, ranging from biomedical applications to energy harvesting, gas sensing, resonant sensing, and timing applications. Ultimately, MEMS sensors and resonators will facilitate the creation of low-cost, small form factor sensors and devices that can be used in a variety of environments and applications, yielding advances in systems such as highly integrated low-power sensors. These new sensors will have an overarching applicability and impact on sectors such as transportation, healthcare, environment, and industrial processes. MEMS sensors and resonators represent an important vector in harnessing the infinite possibilities of the infinitely small.

Frederic Nabki
Special Issue Editor

Review

Silicon-Based Sensors for Biomedical Applications: A Review

Yongzhao Xu [1], Xiduo Hu [1], Sudip Kundu [2], Anindya Nag [3,*], Nasrin Afsarimanesh [4], Samta Sapra [4], Subhas Chandra Mukhopadhyay [4] and Tao Han [3]

[1] School of Electronic Engineering, Dongguan University of Technology, Dongguan 523808, China
[2] CSIR-Central Mechanical Engineering Research Institute, Durgapur, West Bengal 713209, India
[3] DGUT-CNAM Institute, Dongguan University of Technology, Dongguan 523106, China
[4] School of Engineering, Macquarie University, Sydney 2109, Australia
* Correspondence: anindya1991@gmail.com

Received: 10 June 2019; Accepted: 27 June 2019; Published: 1 July 2019

Abstract: The paper highlights some of the significant works done in the field of medical and biomedical sensing using silicon-based technology. The use of silicon sensors is one of the pivotal and prolonged techniques employed in a range of healthcare, industrial and environmental applications by virtue of its distinct advantages over other counterparts in Microelectromechanical systems (MEMS) technology. Among them, the sensors for biomedical applications are one of the most significant ones, which not only assist in improving the quality of human life but also help in the field of microfabrication by imparting knowledge about how to develop enhanced multifunctional sensing prototypes. The paper emphasises the use of silicon, in different forms, to fabricate electrodes and substrates for the sensors that are to be used for biomedical sensing. The electrical conductivity and the mechanical flexibility of silicon vary to a large extent depending on its use in developing prototypes. The article also explains some of the bottlenecks that need to be dealt with in the current scenario, along with some possible remedies. Finally, a brief market survey is given to estimate a probable increase in the usage of silicon in developing a variety of biomedical prototypes in the upcoming years.

Keywords: silicon; sensors; biomedical; semiconducting; nanowire

---

1. Introduction

One of the cornerstones in the field of electronics has been the employment of sensors for the ubiquitous monitoring of different applications. Among the different raw materials being processed by researchers to fabricate the sensors, MEMS-based sensors [1,2] have been the most important. The earliest form of MEMS sensors, which were used to monitor day-to-day applications, dates back to the late 80s and early 90s [3]. With time, the concept of Microelectromechanical Systems (MEMS) has developed in a much easier manner to create a miniaturised version of the sensing prototypes using the microfabrication technique. The usage of MEMS-based sensors has, with time, broadened to a wide range of applications [4–6], including different kinds of alloys and pure material [7–9]. Among the different kinds of MEMS-based sensors, silicon sensors have always been crucial [10,11] for quick and efficient sensing purposes. In comparison to other MEMS-based sensors, the advantages of the silicon sensors involve their small size, excellent signal-to-noise ratio, low hysteresis, ability to work in extreme environmental conditions, and high repeatability in their fabrication structure and responses. One of the significant characteristics of the silicon sensors is their higher response to the changes in frequency in comparison to their enlarged counterparts. These sensors are semiconducting where the substrates are developed from single-crystal silicon. These single-crystal silicon sensors have been developed, characterised and employed to a large extent depending on their type and working principle. The advantages of semiconducting prototypes for sensing purposes lie in the simplicity of their structure and working principle, dynamic nature suitable for implementation, low cost, and scalability. The fabrication of these silicon wafers involves nine stages before they are sent for the fabrication of sensing prototypes—these nine stages are: ingots, peripheral grinding, slicing, beveling, lapping, etching, heat treatment, polishing, and cleaning [12]. The silicon sensors have been implemented in a wide range of applications: ranging from personalised use like healthcare to industrial uses. Among all the applications, the use of silicon sensors for biomedical application has been one of the major sectors where they have, for a while, been creating a major impact [13–15]. The potentiality of these sensors has hence been increased by integrating them with embedded electronics to serve multifunctional medical applications. These silicon sensors are processed by sophisticated micromachining processes to form substrate-like material to be subsequently processed for specific applications. One of the common techniques for processing silicon substrates is to undergo photolithography [16,17] to form the electrode designs on the substrates. This photolithographic technique is further combined with other methods, like etching, thermal oxidation, spin-coating, and sputtering, to develop the sensing prototypes. Some of the arguments in favour of using silicon sensors for biomedical sensing comply with their miniaturised size, high reliability, quick response, lightweight nature, biocompatibility, and minimal hysteresis in their responses. Among the wide sector of biomedical sensing applications that the silicon sensors have been employed for, some of them are applied to physiological body movements [18], human bio-signals like heartbeats [19], blood flow [20], pulse rate [21], drug delivery [22], protein [23] and tumour detection [24], DNA sensing [25], and stem cell research [26]. The performance of the sensors, however, depends on the efficiency in terms of the sensitivity, resistance toward the change in temperature and humidity, robustness, and the enduring nature of these. The biggest advantage, in terms of the connection and trans-reception of data, is the ability of the silicon sensors to be easily attached to the entire signal conditioning circuitry to form the integrated circuitry (IC). Table 1 shows a comparative study between some of the silicon sensors used in different biomedical sensors, along with their strengths and drawbacks. Different kinds of silicon sensors have been used for medical sensing. Some of them are based on piezoelectricity, piezoresistivity, electromagnetic and capacitive sensors principles. The working principle of the sensors depends on the applications for which the sensors are being used.

The researchers have been trying to focus on the adoption of the silicon-based sensing systems for ubiquitous monitoring and to possibly determine the anomaly related to some of the health problems. Although a lot of work review papers have been written [27–30] on silicon-based sensing, there are still some loopholes with the review process that need to be addressed. First, none of the papers has especially addressed an exclusive use of MEMS-based sensors for biomedical sensing. All these papers have considered more than one application, without further elaborating on every aspect. Some of the papers, like [31], have only detailed the work on similar applications of biomedical sensing, like the use of microsurgical tools and microfluidic sensors. Second, among the different types of MEMS sensors, the applicability of silicon sensors, specifically to sense different parameters, has not been described yet. Third, even though some papers [32] have partially worked on the MEMS and silicon-based technology, explaining some of the wireless sensing systems, they have primarily focused on the communication part of the integrated system, rather than on the sensing operation. Some papers have considered the explanation of the significance of silicon-based sensors for certain diseases [33,34], without explaining the need for sensors. Consequently, it is state-of-the-art to elucidate a review paper highlighting some of the significant research work done on silicon-based sensors, which have been particularly used for medical sensors. The use of silicon sensors has to be explained in terms of their fabrication and implementation for a certain aspect of the health parameter. This paper highlights some of the significant work that involves the silicon-based sensors for monitoring biomedical applications. The novelty of this work lies in the explanation of some of the significant work done by the silicon sensors for a wide range of biomedical sensing parameters.

Table 1. A comparative study of the different silicon-based sensors used for biomedical sensing.

Materials	Technique of Fabrication	Application	Strengths	Ref.
Silicon-based MEMS Electric Condenser Microphone	Semiconducting production processing	Human pulse detection	Smaller size, better quality than other ECMs	[21]
Silicon Nanowire	Bottom-up approach	Detection of DNA molecules	Thermally and chemically stable, interconnects better with the components	[35]
Silicon probe, PEDOT: PSS, polyimide	Monolithic microfabrication process	Detection of neural activity	Implants several probes in the brain within a short time	[36]
Silicon-based CMOS and BiCMOS	Photolithography and chemical process	Heartbeat and respiration activity	Wireless communication, high data transfer rate	[37]
Amorphous silicon-image sensor based on thin-film transistors	Thin-film semiconducting process	X-ray Diagnostic Medical Imaging	Low data lines capacitance, noise cancellation techniques and optimised timing	[38]
Silicon-based CMOS and BiCMOS	Photolithography and chemical process	Detection of peripheral and cranial nerve activities	Enhanced biological and electrical performance of the implantable sensors	[39]
Silicon-Silicon dioxide-Chromium	Conventional photolithography process	Detection of proteins and photo lipids	Reduced electrode impedance, higher sensitivity, reduced dependence on cell mobility	[40]
Nitrogen-doped silicon	Thermal oxidation and deposition	Detection of protein (Avdin)	Low detection limit and high sensitivity	[41]

The manuscript has been sub-divided into four sections. The introduction given in the first section showcasing the importance of silicon-based sensors for biomedical sensing is followed by Section two, which explains some of the works done on the detection of health parameters using silicon sensors and the corresponding embedded systems. This section presents seven different types of work, based on the nature of the use of silicon for biomedical sensing. The third section elaborates about the challenges related to some of the existing sensors, as well as some of the possible remedies that can be associated with them. It also provides a glimpse into future opportunities in terms of market surveys on silicon-based sensors. The conclusion of the paper is presented in the final section of the manuscript.

2. Utilisation of Silicon for Biomedical Applications

The proposition for the use of silicon for medical applications has existed for almost four decades now [42,43]. Since then, silicon has been used in different forms for a range of in-vitro and in-vivo applications. The different forms of silicon have been mechanically and electrically enhanced via microfabrication technologies. For the last three decades, silicon-based sensors have found significant applications in industry and medicine [44–46]. The origin of silicon-based sensors can be found in 1954, when Smith et al. for the first time introduced the term "piezoresistivity" by studying the stress sensitive effects in silicon along with germanium [47]. During the early 1960s, the first silicon pressure sensors and strain gauges were developed and reported from the Bell labs and the Honeywell Research Centre [48]. During these years, silicon sensor technologies had become quite popular, and by the late 1960s, different US companies had already produced the first silicon pressure sensors. Furthermore, the combination of silicon technology with information and communication technologies enabled the development of compact, low-cost and high-performance devices for different applications [49–52].

Nowadays, biological and biomedical silicon-based technology has exhibited a remarkable potential in the application field, from a research point of view as well as an industrial perspective [53,54].

2.1. Planar Sensors

As to the use of silicon-based prototypes as planar sensors, researchers have used them for electrochemical sensing for biomedical applications [55]. Due to their structural advantages, like the variation of the distance between the two electrodes in order to simultaneously vary their electrical characteristics, planar sensors possess certain advantages like a lower input power, higher efficiency and robustness, and higher ionic diffusion. Consequently, these sensors are highly efficient for developing a lab-on-a-chip system for biomedical applications. Due to their reduced scanning area for the nanoscaled electrodes, the sensitivity is higher when compared to other electrode patterns. In planar capacitive sensors, the sensor electrodes are positioned in a coplanar surface. In addition to the mentioned features, the planar configuration offers the possibility of evaluating a material under test (MUT) from one side only [56], which is especially useful when the access to MUT is restricted. These characteristics make capacitive sensors a vital device for applications, such as non-destructive testing (NDT) [57], proximity and displacement measurement [58,59], material characterisation [60], and imaging [61,62]. Planar capacitive sensors operate based on parallel-plate capacitors. The electric field lines bulge from one to another due to the planar nature of the sensors. This leads to the formation of fringing electric fields. When the electrodes open to a coplanar plane, the fringing electric field becomes the largest between the working and sensing electrodes [63]. The fabrication of the sensors required a thin-film technology, along with photolithography and etching techniques to form the respective substrates and electrodes. Some of the substrates used to form the sensors were silicon nitrate, silicon dioxide, polycrystalline silicon and aluminium. The thin-films were formed using the lift-off technique through a stencil, followed by the evaporation process to deposit the metal on the wafer. Then, the stripping process was carried out for the operation metal to obtain the contacts of the sensors at defined patterns. Finally, sputtering and vacuum evaporation was carried out to define the electrodes of the sensors. The electrodes were developed in the form of interdigital capacitive sensors for certain advantages, like a small voltage drop, small charging currents, and a fast steady-state mass transfer. The dimension of the microelectrodes was less than 20 µm, where the space between two consecutive electrodes is in single micro units. The thickness of the metallic layers on these interdigitated electrodes was around 250 nm for polycrystalline and monocrystalline silicon. The detection of urea, DNA and other proteins was done using these sensors via conductimetric measurements. Musayav et al. [64] developed a microarray sensor for detecting the direct phosphate backbone charge of DNA molecules. Capacitive metallic electrodes with dimensions of 7 µm × 7 µm had been utilised to develop the array by arranging the electrode with a pitch of 15 µm. Since the sensor was used for a low-noise sensing capability and amplification circuits, along with double sampling, the sensitivity has been quite improved, and it was found capable of sensing DNA having a 1 pM concentration. This research group also noticed that surface treatment is required to improve the sensitivity, and, therefore, the gold coating was retained for future work. Capacitive sensors have been employed in some applications because of their low cost, quick response, as well as their non-invasive and no radiation features in electrode design [63,65].

Afsarimanesh et al. [66] have also used silicon-based planar sensors to determine the change in concentrations of CTX-1, which is one of the biomarkers used to study the condition of the bone. The change in the concentration of CTX-1, which is one of the fundamental proteins, has been studied by the researchers to determine the amount of resorption of bone. Specific concentrations of CTX-1 ranging between 0.1 to 2.5 ppb were produced using a serial dilution to determine the capability of the silicon sensors to measure them. This range of concentration was decided based on the amount of CTX-1 being released from a patient who had osteoporosis. Electrochemical Impedance Spectroscopy (EIS) was used to study the response of the sensors in terms of the change in impedance with respect to the frequency. The silicon sensors fabricated with the photolithographic technique were coated with a polymeric layer, which consisted of templates specific to the CTX-1 molecule. The electrodes were interdigital, operating on the planar capacitive principle. The sensing area of the sensors was 2.5 mm × 2.5 mm. The molecularly imprinted polymer was formed to coat a colloidal layer on top of the sensing surface to make the sensor sensitive toward the particle CTX-1 molecule. The sensitivity of the sensors was high, with a quick response of less than one second. The sensor was integrated with a microcontroller-based system to make it portable for real-time applications. The development of the embedded system signifies some other qualities of the sensors. Some of these include the quick data-collection process and testing of samples from different patients at a very short time. Other advantages of this system are its low-cost, ease of operation, robustness, and a very low detection limit.

2.2. Polysilicon-Based Sensors

The next category is the use of polysilicon to develop the sensing parts of the prototypes. The advantages of using polysilicon lie in the controllability of the thickness of the formed prototypes, their durability within high changes in temperature, ability to tune in the resistance to simultaneously change the temperature coefficient of the resistance, and compatibility to form hybridised prototypes in conjugation with other metals. Fernandez et al. reported a triglyceride measurement system [67], which was made of composite porous silicon/polysilicon micro-cantilevers. Micro-cantilevers can transduce different types of chemical and physical occurrences into mechanical motion. Micro-cantilevers have been used in a range of applications, such as those required for sensing cells, proteins, and metals. Arrays of these cantilevers can detect multiple factors at once. This biosensor has a cantilever beam with a length of 100–200 μm, width of 10–20 μm, and a thickness of 2 μm. Abdelghani et al. [68] fabricated a silicon-based capacitive pressure sensor to measure the blood pressure and heart rate by studying the deflection of the diaphragm resulting from the applied pressure. The pressure causes deformation on the movable upper plate and creates a gap between the upper and the fixed lower plate, which generates the capacitance. This research group also established the FEM model to analyse the deformation of the membranes. Due to its low power consumption and advantages resulting from miniaturisation, it has a great capacitive response within the pressure range of 0–40 kPa.

2.3. D printed and Optical Sensors

With the advancement in the field of 3D printing, researchers have begun to emphasise the development of 3D printed silicon-based sensors for biomedical applications. These sensors are quick to fabricate, able to be customised in a wide range, robust in nature and highly durable. One of the significant issues with these materials is their biocompatibility, especially for the sensors that are used for in-vivo applications. The significance of these 3D sensors lies in their integration with other organic and metallic elements to form sensing prototypes [69]. These sensing systems have an additional attribute of transmitting the monitored data wirelessly for smart health monitoring. Other characteristics of these sensors are their biodegradable nature and their stability in their responses for over three months. The sensors are operated as battery-less capacitive and inductive pressure sensing prototypes at high resonating frequencies of 130 and 183 MHz. The sensors have been used to measure the change in blood pressure of different animals. The fabrication process started with the deposition of a copper layer of around 200 nm on top of the 300 nm thick oxidised silicon layer. Electroplating was done on the copper layer to form moulds to define the inductor coils. This was followed by photolithography to form the patterns on the substrates. Then, the SU-8 layer was used to form the topmost part of the sensor with a thickness of 100 μm. Then, the oxidised silicon layer below the SU-8 layer was removed by a buffer solution, followed by forming a bottom layer of SU-8. Finally, electrodes were formed with Ti and Au with a thickness of 10 nm and 100 nm, respectively. The sensitivity of the measured pressure range for these sensors was around 160 kHz/mmHg, with an error range of 2%. The sensors were found capable of determining the change in phase angles with respect to a frequency range between 100–180 MHz for different movements of the attached antenna. The validity of the biocompatibility and the experimental results from these sensors were obtained by inserting these sensors in three different locations in the body of a mouse, as shown in Figure 1 [69]. The sensors could successfully measure the blood pressure of the animal, ranging between 91–129 mmHg. The robustness and repeatability in the responses were also validated by shifting the resonating frequency within a range of 3 to 7 MHz.

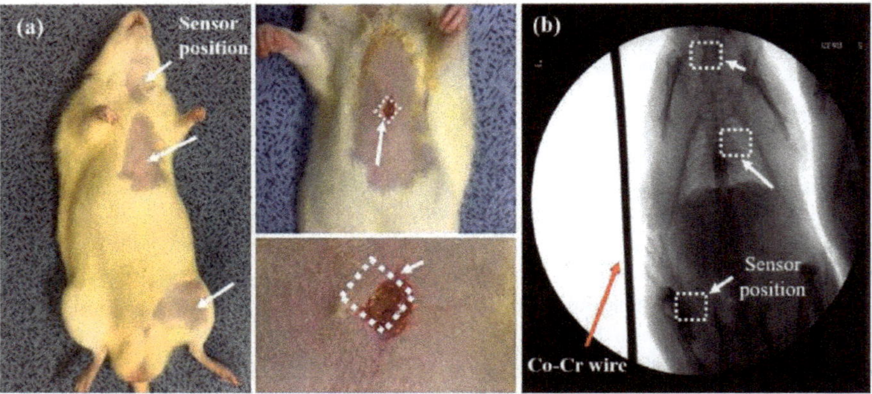

Figure 1. The insertion of wireless pressure sensors into the animal was done (**a**) at three different places (**b**) to perform the biocompatibility test and determine the blood pressure [69]. The image has been reproduced with permission from [69].

Initially, the use of silicon in terms of optical sensors was done by developing a fibre-optic pressure sensor [70] for biomedical applications. In addition to their small size, their resistance toward electromagnetic radiation and their ability to monitor at remote distances make them a popular choice for biomedical applications. The dimensions of the sensors were 270 µm × 270 µm × 150 µm. The working of the sensors is based on the principle of detection of the intensity of reflection of light from a diaphragm. The biomedical use of these sensors is based on balloon catheters. The catheters were formed with Polyethylene terephthalate (PET) and had a diameter of 1.5 mm. Anisotropic etching and direct wafer bonding techniques were used to fabricate the sensor. The sensing structure consisted of Gold and Chromium thin-films, along with a silicon-based optical fibre-stopper and fibre-aligned structures. The fibre-stopper was formed using a sodium silicate solution, whereas the alignment was done between the bottom and middle silicon wafers by pressing with a rubber roller. The sensors were annealed at 200 °C for two hours to achieve a proper attachment and alignment between the two layers of the sensors. The incident light was from an LED having a wavelength of 1.31 µm. The optimisation was done on the area and thickness of the working environment to operate on a pressure range of 0.1–1 MPa. The response of these sensors to such a high range of pressure also increases their significance for a wide sector of applications operating at temperatures at middle and high ranges. The pressure was achieved using a dry nitrogen chamber to determine the intensity of the reflection with respect to the variation of the pressure. The sensors obtained an average responsiveness of 1.9 µmW MPa^{-1}.

2.4. Ion-Sensitive Field-Effect Transistors

Sensors were also developed in the form of Field-Effect Transistors (FETs) and Ion-sensitive Field-Effect Transistors (ISFETs), where the gate is made sensitive to particular ions, such as Na$^+$, NH$_4^+$, K$^+$ and others [71,72]. These types of sensors have gained a significant importance due to their employability in biomedical applications, which results from their ability to measure different kinds of analytes, swiftness of response, high robustness, and low output impedance. The techniques to fabricate FETs and ISFETs have been consistent over the years, with their resultant sizes decreasing every day to increase their efficiency and sensitivity. This performance can also vary commensurate to the level of doping done on every individual sensor. The sensors were also used for bio-sensing to measure certain elements, such as a2–interferon [73] and β–Bungarotoxin [74]. Here, the sensors were able to detect the toxin present in nano levels. One of the earliest works related to the use of silicon for biomedical sensing has been on the employment of silicon needles [75] to develop ion-sensitive field effect transistors (ISFET).

The sensors consisted of multi-sensing silicon needles, including a pseudo-reference electrode developed from platinum and a temperature sensor. The electrode and temperature sensor were developed via CMOS-compatible technology and a silicon micromachining technique. The sensors displayed a high sensitivity and good linearity toward the detection of myocardial ischemia during cardiac surgery. The applications in cardiac surgery with these sensors were determined by the change in pH and pK. P-type silicon wafers were used with a resistivity of 4–40 Ω cm, having a high doping level of 1×10^{15} cm^{-3}. The ISFETs were fabricated via a photolithography technique, at six different levels, having a gate and LPCVD of silicon dioxide and silicon nitride, respectively. The shape of the silicon needles was defined through a reactive ion etching process, which was independent of the crystalline planes of the silicon surface. The length and width of the needle were around 7 mm and 0.8 mm, respectively. The sensors were equipped with a membrane of plasticised potassium, which makes them highly sensitive towards potassium. The sensors showed a change toward the pH and temperature, obtaining a range of linear responses between 8×10^{-5} and 8×10^{-2} M and a sensitivity of 50 ± 2 mV per electrode. The response time and temperature coefficient of the resistance of the ISFETs were one second and 2687 ppm/C.

2.5. Silicon-On-Insulator Sensors

Another work based on silicon-based sensors for biomedical sensing from a different sector was the employment of silicon as insulators to form wireless implantable sensors [76]. These sensors were used for fully implantable cochlear prosthesis and, brain sensing technology, consisting of an embedded signal-processing circuit. These systems hold a key role for the human prosthetic and brain-machine interfacing systems. The system was fabricated based on a silicon-on-insulator (SOI) technique to design the proposed architecture. Some of the advantages of this system include the simple design, common-mode interference rejection, low noise, null DC power dissipation, and straight forward interfacing between the sensor and the system. One of the matters under consideration was that the acceleration noise floor limited the resolution of the device. This phenomenon was due to the Brownian noise, which was directly proportional to the resonant frequency of the mechanical device, and inversely proportional to the mass and quality factor. The accelerometer consisted of a set of sensing fingers having a thickness of 25 μm, width of 2 μm and an overlap dimension of 96 μm. The sensing surface of the system was around one mm^2, while the entire system had a dimension of 2.5 mm $\times$ 6.2 mm, obtaining a capacitance of 2 pF. The sensitivity and the resonant frequency obtained by the system were 11.5 mV/g and 6.4 kHz. For the cochlear prosthesis measurements, the system was able to detect 60 dB SPL, 35 dB SPL and 57 dB SPL at 500 Hz, 2 kHz and 8 kHz, respectively. For the brain-sensing technology, the sensors were able to distinguish between the on and off states for a spectrogram low-pass filter recording of a person with Parkinson's disease. The measurements were taken within the chosen beta band between 10 and 20 Hz. Local field potential signals were recorded at low frequencies of around 200 Hz to reduce the generated noise. The Chopper-stabilization mode was employed to determine the noise performance for the operation of the system, in order to improve the common-mode rejection ratio. The stability and the biocompatibility of the sensing system were ensured by the size and type of the chosen materials. Commercially available macro electrodes, developed from an alloy having an impedance of around one kΩ, were chosen to form the conductive parts of the sensors. The response time of the sensors to detect interictal epileptiform singles was more than one second. Even though the developed sensors were operated over a wide range of frequency, one of the disadvantages of these types of sensors was found to be their high output impedance, which increases the power required to drive the system. The sensors were also not as small as the other types, which eventually decreased their sensitivity. One of the other points on which the researchers can focus in this work is the reduction of Brownian noise generated in the system. Other works on Silicon-on-Insulator wafers were conducted using the coating on PZT thin films while fabricating standard piezoelectric MEMS microfabrication technology that generated ultra-thin PZT/Si structures [77,78].

2.6. Silicon Nanowires

There has been a tremendous increase in the use of nanowires because of their significant advantages, such as a high efficiency, high selectivity and sensitivity. Silicon nanowires have been developed along the same lines, keeping in mind the aforementioned advantages, and hence served the same purpose. Silicon nanowires hold a huge potential as conductive materials due to their widely accepted attributes. Some of them include a high electrostatic controllability [79] with better current characteristics and a lower noise density. Furthermore, in comparison to metallic nanowires like iron, silicon nanowires are relatively easier to fabricate. This is due to the low-pressure operation, the feasibility of the fabrication of these nanowires in batches and the low-cost setup. Another advantage of these sensors is their good orientation, unlike certain nanoparticles like carbon nanotubes, where the orientation is effected by the tunneling current that exists between them [80]. Silicon nanowires have shown the potential to be used for both electrochemical and pressure sensing applications in the field of biomedical sensing. The range of analyte being detected with the help of these nanowires is also exceptional. Chen et al. [81] explained the use of silicon nanowires in developing field-effect transistors for the detection of proteins, DNA sequences, small molecules, and other biomarkers for certain diseases like cancer. The physiological responses of the SiNWs-FETs from the cells and tissues were also measured. The fabrication and experimental processes are shown in Figure 2 [81]. The sensing system consisted of three-electrodes, namely the source, drain, and gate electrodes, which formed a bridge on the semiconducting channel. The channel was developed from SiNW, which was responsible for the attachment of bio-receptors via a chemical modification. The attachment of the receptors was completed to recognise the specific target analyte, in order to achieve a high sensitivity, specificity and strong affinity. A detection system was used to determine the variation in the conductance and surface potential of the channel during the interaction between the receptor and target. Two different techniques, namely top-down and bottom-up, were used to fabricate the silicon nanowires on oxidised silicon wafers with a thickness of around 200–400 nm. The nanowires and the connecting electrodes of the SiNW-FETs were formed on the sensors using a range of techniques like photolithography, ion-implantation, reactive ion etching (RIE), electron-beam lithography, and thermal evaporation. With this approach, the width of the nanowires was around 100 nm. The silicon substrates were doped with boron while the nanowires were doped with different elements like nitrogen and phosphorous to alter their semiconducting property. The bottom-up approach was done through a chemical vapour deposition process, which was followed by the assembly of the formed nanowires, using an electron-beam lithographic process. The advantage of this process is associated with metallic nanoparticles having controlled sizes with the help of a catalytic activity. Then, the nanowires were assembled using techniques like flow-assisted alignment, the Langmuir-Blodgett process, being bubble-blown, smearing-transfer, roll-printing, and PDMS-transfer. After the dispersion of the nanowires on the oxidised silicon wafers in the central areas to form the channels, they were subsequently connected with metallic electrodes. The electrodes were then coated with an insulating layer to minimise the leakage of the current during the experiments. The sensor consisted of a sensing area of around 1.5 mm × 1.5 mm containing the FET array, a microfluidic channel and a detection system. The sensing system was finally mounted on a plastic circuit board having electrical connections built with aluminium wires of a diameter of 30 μm. During the experimental process, the FETs were immersed in acidic buffer solutions, utilising the high surface-to-volume aspect ratio of the nanowires. This affected the external electric field, which subsequently changed the resultant conductance of the devices. Some of the biomedical experiments carried out with these devices were protein-protein interactions, DNA hybridisation, the monitoring of viral infection, peptide-small molecule interactions, and biomarker detection. The corrosion of the source and drain electrodes was prevented during the experimental process in harsh environments. The FET-probe of the sensors had a sensitivity of 4–8 µS/V for the tested solutions.

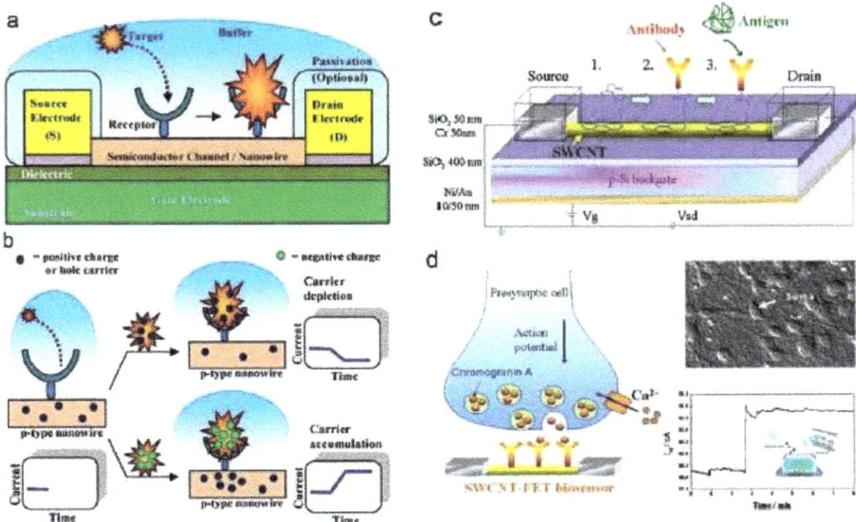

Figure 2. (**a**) Schematic diagram of the nanoscaled-FET silicon-based sensors. (**b**) Binding of the target molecules with the receptors. (**c**) Surface modification of the FET device done with a π-π interaction process, immobilisation and detection. (**d**) Release of the neurons on the sensing surface of the FETs to determine the change in current with respect to the time [81]. The image has been reproduced with permission from [81].

Another use of silicon nanowires has been shown by Abdolahad et al. [82] by developing a biosensor for the electrical charge sensing of cell membranes. Silicon nanowires, with the help of a gold catalyst layer, were fabricated in a low-pressure chemical vapour deposition (LPCVD) system for the electrical monitoring of the negative charges that are present on the membrane. It has been observed that during the detection of the negative particles on the surface of the cancer cell membrane, the primary colon cancer discharges more current, confirming progressive colon cancer. Thus, by differentiating between the electrical responses received from the device according to the different negative charge productions, cancer cells can be differentiated or categorised. The silicon object is coated with a 10 nm gold thin catalyst layer on which the specimen needs to be positioned inside the LPCVD system. This SiNW bioFET can be very useful as a diagnostic tool for cancer examination in future biomedical applications. Figure 3 depicts the schematic diagram of the FET fabrication process [82]. Another similar study had been done by Zhang et al. [83] for DNA sensing using field-effect based silicon nanowire (SiNW) sensors. Since the gap between the DNA charge layer and the SiNW surface is highly dependent on the sensitivity, in terms of the ionic strength of the electrolyte solution, it becomes weaker when the gap is further increased. This depends on the hybridisation spots of DNAs to the peptide nucleic acid (PNA) capture probes. Polysilicon layers had been covered on the thermally grown oxide on a p-type test wafer in order to design this biosensor with a dimension of a few tens of a nanometre.

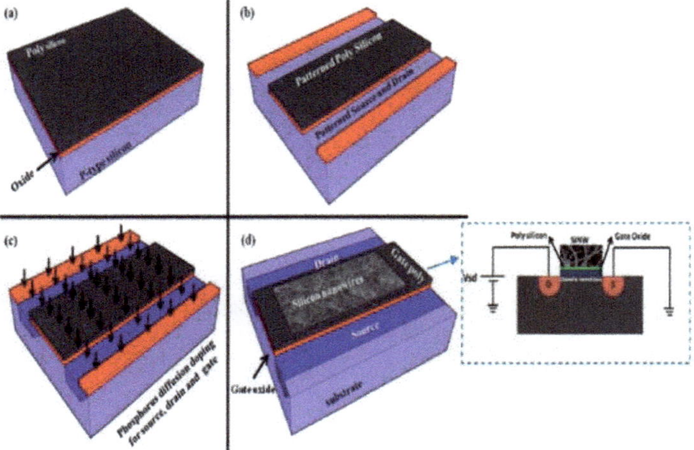

Figure 3. The schematic representation of the skein SiNW incorporated FET fabrication process. [82]. The image has been reproduced with permission from [82].

Carrara et al. have been working for a while on the fabrication of memristors with silicon nanowires [84] and their further implementation for biosensing purposes [85,86]. A combination of the Complementary Metal Oxide Semiconductor (CMOS) processing technique and photolithography was used to develop poly-crystalline silicon nanowires that exhibited memristive behaviour. After developing an oxidised layer of silicon on the silicon substrate, a thin film of polysilicon was developed by LPCVD and deposited with a thickness ranging between 40 and 90 nm. This was followed by the etching process to remove the horizontal layer. A spacer was defined, which was reminiscent of the nanowires. The nanowires were polysilicon in nature, having thicknesses from 20 to 60 nm. The contact regions of the nanowires were developed with chromium and nichrome. These devices displayed an ambipolar behaviour with an increase in the conduction of the current due to the presence of holes. For the computational study, their behaviour was studied under a certain frequency range to determine the changes in the current with respect to the voltage. The silicon nanowires were treated as resistors, being placed in FETs with a Schottky source and drain contacts. The sensors were tested with three antigen concentrations, which showed a rise in the capacitance with respect to the voltage gap. The concentrations of antigen varied between 0 and ten fM. The gap was created due to the occurrence of the functionalization of the antigen with the antibodies. This change occurred due to the change in the Schottky barrier as a result of the change in the electric potential during the injection of the carriers. This study proved that the behaviour of the physical system was in accordance with the bio-modified surface as a result of the changes in the capacitance of the nanowires.

The testing of these nanowire-based memristive devices was also done for the prostate-specific antigen (PSA) IgM detection. A label-free detection of the biomarker was performed, where three low concentrations of the PSA were tested and analysed. These experiments were conducted to validate the performance of these devices for the early detection of prostate cancer disease. The results were cross-checked with an enzyme-linked immunosorbent assay (ELISA) technique to further determine their accuracy. The substrates of the silicon nanowires were functionalised with different antibody solutions via an over-night incubation. The concentration of the PSA-IgM was varied between 11.75 and 47 AU/mL, with the precision being maintained by removing the excess of the solution using 1.5 mL of PBS-Tween 0.05% and MilliQ water. The drain to source voltage was swept between −2.4 V and 2.4 V to determine the changes in the current under an environment with a controlled current and humidity. A hysteretic loop was obtained for the current-to-voltage characteristics, with a voltage gap being created after the bio-modification of the nanowires. The charged residues of the proteins changed the characteristics of the electric field, which subsequently changed the conductance of the device.

2.7. Piezoresistive Sensors

The design of each of the silicon-based sensors varied with the type of application they were being used for. In addition to the traditional piezoresistive sensors developed by different polymeric materials, silicon-based piezoresistive sensors have served a better purpose in this respect as they have been advantageous for a wide range of pressure-based biomedical applications. Some of the advantages of these sensors are their improved accuracy, expanded pressure range, lower power consumption and reduced cost of production. Beccai et al. [87] explained the development of a hybrid three-axial sensor to force sensing in biomechanical applications. The sensors were developed on a 525 μm wafer having a high aspect-ratio cross-shaped flexible element. The piezoresistive working principle was followed by these sensors, with four resistors operating in a combined fashion, each one having a range of length and width of 30–50 μm and 6–10 μm, respectively. A hybridised integrative approach was adopted for the interconnection and operation of these force sensors to achieve a high linearity and low hysteresis for both normal and shear forces. The sensor was designed to have a combined connection of carrier chip and flip-chip technology. The final sensor had dimensions of 2.3 mm × 2.3 mm × 1.3 mm. The sensors were used to determine the change in the relative resistance concerning the different forces applied to them.

One of the studies highlighting the use of silicon-based sensors for neural studies was done by Manikandan et al. [88] by designing and developing intracranial pressure sensors. The design of these sensors was based on the multi-electrode array, where the size of each sensor was dependent on the size of the chip and package. The size of the total chip was around 100×200 μm^2, where each of the sensors operated based on a piezoresistive effect. The total sensing chip also consisted of a temperature and a separate pressure sensor. The resistance measured from the sensor array was based on the Wheatstone bridge to operate the sensors as passive resistive resonant prototypes, in order to accurately determine changes with respect to intracranial pressure. The sensitivity of the system was around 0.84×10^{-2} mV/kPa for a maximum pressure limit of 112 Pa. The sensors consisted of four round-shaped microelectrodes developed from Silver and Platinum. The entire array was divided into three sides, consisting of simulation sites, electro-physiological sites, and a pressure sensing side. There were some other sites that were kept open in the sensor array for the detection of certain proteins like glutamate oxide, from a matrix layer. The connecting pads for each of the two electrodes had a size of around 500 μm × 500 μm for the transfer of the monitored signals.

The two electrodes were used as counter and reference parts, where the difference in the voltage signals between them was measured by the embedded detecting circuit. The fabrication of the sensor was done on silicon substrates, with a thin layer of polyimide having a thickness of 15 μm being coated to form the base for the electrodes and temperature sensor. After the electrodes were formed using Au and Pt, another polyimide layer was coated on the surface. This was followed by the defining of the electrodes by performing a reactive ion etching technique on the top layer to form the openings. The resistive temperature sensors were formed by sputtering Silver under temperature and vacuum conditions of 350 °C and 1×10^{-6} mbar, respectively. The electrodes were positioned at specific locations to determine the pressure of the neurons without the assistance of any special gel. In the presence of the four sensors, two of them were calibrated to measure the variation of the pressure. The measurement of the temperature was done between the range of 0 and 50 °C to calibrate the sensor and determine the linearity of their responses. The sensing chip was able to respond between the range of −20 °C and 20 °C, having an accuracy of around 89% in the response. The effect of the external pressure on the piezoresistive sensors was also determined with the change in resistance.

One of the earlier works related to the use of silicon-based piezoresistive sensing was done by designing and developing pressure sensors for biomedical applications [89]. The sensors were fabricated to respond to low pressure to achieve an enhanced performance factor that was defined by the product of the signal-to-ratio, sensor sensitivity and temperature coefficient of the piezoresistance. The sensors achieved the highest figure of merit for the sensors having a dimension and thickness of 100 µm and 10 µm, respectively. The optimisation was done based on a mathematical calculation to determine the output voltage of the sensor, sensitivity, noise, and signal-to-noise ratio. The doping concentration of the sensors was inversely proportional to their sensitivity and directly proportional to the signal to noise ratio. The coefficient of the temperature initially increased with the doping concentration and then decreased after reaching a certain value. The optimised values of the thickness of the diaphragm, doping concentration, and length of the sensor were 15 µm, 5×10^{17} atoms cc^{-1} and 25 µm, respectively.

Despite the reduced size of these sensors, they could also respond to a varied range of temperatures at a high sensitivity and pressure limit. The fabrication process of the sensors is also compatible, as the sensing area was developed using standard MEMS-based techniques. One of the factors that is instrumental in determining the performances of these sensors is the doping level, which significantly varies the response to a range of pressures, thus affecting its sensitivity.

2.8. Integrated/Hybrid Sensors

The last category of silicon-based sensors is based on the formation of different kinds of sensing elements for multifunctional biomedical applications. These sensors are different from other types, as explained above in terms of the fabrication technique and the resultant prototypes. Another significant attribute of these types of sensors is their customised nature, which increases the range of applications they can be employed for. Instead of following a specific working principle specialised in the application type, integrated sensors were found to operate on more than one kind of principle. One study explains the silicon-based blood pressure sensor designed by Wu et al. [90] using the fractional flow reserve (FFR) technique. The diameter of the sensor is 125 µm, and a reference manometer was used to compare the pressure results. Due to the examining capability of the location of a stenosis in the coronary artery, the fibre optics have a great advantage in the fabrication of this sensor. The use of a silicon dioxide diaphragm of uniform thickness eliminates the bulky structure of the sensor head. During the blood pressure measurement, the inflation/deflation cycle gives an electric voltage signal from which the results of each transient period are received. First, the sensor receives information regarding the heartbeat signal, and then the blood pressure is monitored at different positions in the coronary artery. Satake et al. [91] fabricated a micro silicon-based sensor that can count the number of blood cells in human blood and the latex particles using the Coulter method. It is also possible to distinguish the white and red blood cells via this microsensor. A glass cover attached to a silicon substrate was used to develop this particular sensor, which has a dimension of 10 mm × 5 mm × 0.8 mm. The researchers proved that the presence of linearity between red blood cells and the counts could be measured. The height and the number of pulses help to determine the differences in size and the concentration of PSL particles, respectively. One of the advantageous attributes of these sensors is that they can be fabricated using both glass and silicon. The smaller dimension of the sensors is another advantage to it.

A multifunctional silicon-based sensor was designed by Kang et al. [92]; it was tested by being implanting in the brain of a rat. This bio-resorbable electronic sensor has the wireless ability to monitor the intracranial pressure (ICP) and intracranial temperature (ICT) regularly and simultaneously, which can be very helpful for the treatment of a traumatic injury of the brain. This research group utilised a wireless transmitter-based sensor for the data transmission. The fluid flow and motion, thermal characteristics and pH can also be obtained, which are related to sudden pain in the body when they are installed in deep tissues or body cavities.

This biosensor is not only helpful for the diagnosis of chronic health illnesses such as diabetes, but also very helpful for the treatment of the cardiac space and spinal system by sensing and stimulating. Figure 4 depicts the schematic diagram of the location, on the rat, of the sensors for pressure and temperature sensing using wireless communication [92]. The key characteristics of this sensor include the measurement of multiple intrinsic parameters of the brain, along with the determination of its behaviour for other physio-chemical features.

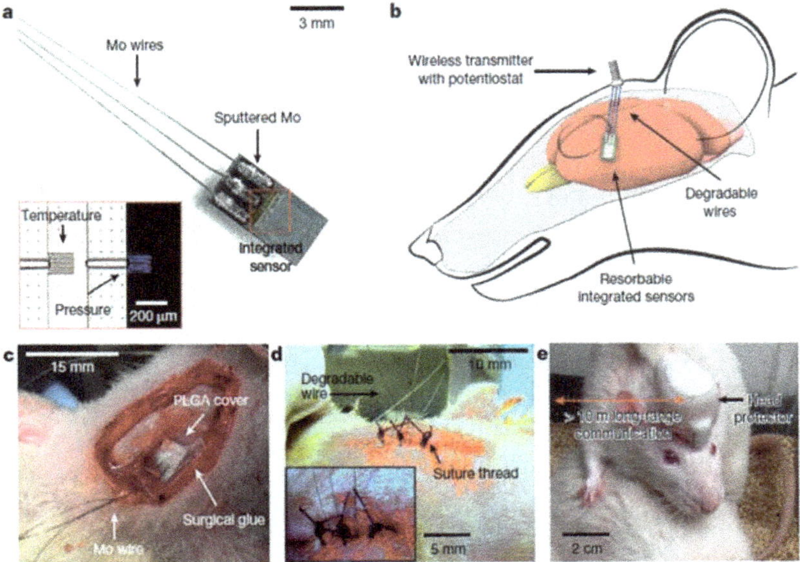

Figure 4. Schematic diagram to represent the interfacing of Bioresorbable sensors with the communication modules for the wireless data transfer. The image also shows the connection of the bioresorbable pressure and temperature sensors integrated with the dissolvable metal interconnect [92]. The image has been reproduced with permission from [92].

Hui et al. [93] have showcased work related to the use of silicon wafers to develop bio-integrated flexible electronic systems. The integration of silicon wafers was done with flexible electronics, the key characteristic of these sensors. The integration of two types of materials led to the achievement of the enhanced electrical conductivity and mechanical flexibility of the sensors. The thickness of the layers grown on the wafers was under 1 µm. The sensors have been proven to be effectively used for chronic implants. Thermal oxidation on silicon wafers was done to grow a thin layer of silicon dioxide; the process was followed by electron beam evaporation, photolithography, and etching to achieve specific electrode patterns. Magnesium was encapsulated by two layers of silicon dioxide and a mixed layer of aluminium oxide and Parylene C. The layer of magnesium with a thickness of 200 nm was formed to examine the behaviour of the sensors toward the water barrier properties. The silicon dioxide layers were then transferred to a thin polyimide film having a thickness of 25 µm. The polyimide film was supported by a temporary glass substrate for manual manipulation purposes. Electrochemical impedance spectroscopy (EIS) was used to detect the performance of the thermal growth of the silicon dioxide layer by developing an equivalent circuit. The circuit consisted of two capacitive elements, namely oxide and double-layer capacitances, and two resistive elements, solution and charge-transfer resistances. Although the testing of these hybrid sensors was done using one element, the response of these sensors toward EIS has opened a huge potential in determining the response of these types of sensors for other analytes, such as proteins, for a range of frequencies. The electrochemical cell responding to the frequencies can be optimised by including different kinds of polymers to conjugate with silicon and other metallic elements.

One of the studies explaining the use of silicon carbide for biomedical applications has been showcased by Gabriel et al. [94] in the form of multi-sensor multi-probes. The use of silicon carbide as a semiconducting substrate has been popular for the last decade due to its capability of working in high-power applications. It was seen that the performance of silicon carbide as a substrate was better than that of silicon from the particular set of experiments that were performed to determine the change in impedance with respect to the frequency for different tissue properties. Initially, a layer of Pt/Ti with a thickness of 450 µm was deposited on silicon and silicon carbide, where the former substrate was oxidised by CVD to form an SiO_2 layer of 1.5 µm. The area of the system was around 14.85×0.83 mm^2 with the presence of two pair of electrodes, each one having an area of 300×300 µm^2. The sensors were mounted on printed circuit boards with gold contacts, followed by the electrodes being platinised. This was done to reduce the total impedance of the sensors. The treatment was done with KCl solutions, along with platinum chloride and lead acetate. The in-vitro measurements were done with a four-electrode sensor via the immersion of the impedance probes in the diluted and non-diluted physiological solution. The frequency sweep was done between 10 Hz to 1 MHz to determine the changes in the impedance and phase angle for both the silicon and silicon carbide substrates. It was found that the silicon carbide substrates outperformed the silicon sensors in terms of the load-deflection tests, in-vitro characterisations, and deflections, and this simulated life-like situations. The importance of silicon carbide operating on a range of frequency sweep can be used to exploit its performance for other in-vitro and in-vivo applications. The use of silicon both as electrodes and substrates does create a uniform structural attribute in terms of the behavioural aspect. These sensors were also smaller in size, which makes them suitable for determining the changes in some of the elements inside the body at minuscule levels.

One of the recent silicon-based works for medical applications has involved the development of bioresorbable pressure sensors for chronic diseases and healing processes [95]. Some of the diseases being dealt with by these pressure sensors are traumatic brain injuries, glaucoma and hypertension, based on the pressure in different parts, like the intracranial, intraocular and intravascular spaces, respectively. The experiments were conducted on rats for 25 days to validate the lifetime and repeatability of the response of the sensors. Apart from the sensors being biodegradable, the results in terms of the processed materials, blood counts, blood chemistry, and magnetic resonance imaging compatibility showcased the clinical usage of the device. Figure 5 shows the schematic diagram of the fabrication steps of the sensors [95]. The sensing system consists of four silicon nanomembranes, each having a thickness of 200 nm. The diaphragm of the system was developed to form a floating, pressure-sensitive region of air-filled cavity, having the dimensions of 200 µm $\times$ 200 µm $\times$ 10 µm. The temperature sensor was located away from the diaphragm, so that the response of this sensor would not get affected by the pressure. The diaphragm consisted of monocrystalline silicon and evaporated silicon dioxide layers of 200 nm and 600 nm, respectively. The electrode layers were developed with silicon nanowires, with overall dimensions of 200 µm $\times$ 200 µm. The sensing system was operated as a Wheatstone bridge consisting of a voltage source, temperature sensor and four pressure sensors. The thickness of the silicon nanowires was optimised to increase the pressure sensitivity of the device. The resistance values of the sensors changed with respect to the temperature in order to obtain a linear response, having a temperature coefficient of 0.0012 °C^{-1}. The sensitivity of the pressure sensors varied within a range of $\pm$ 1.5% for 22 days. The sensors were mounted on a thin film of poly (lactic-co-glycolic acid) and bonded with the surrounding skull with glue. The outputs of the sensors were measured in terms of the change in voltage with respect to the time for different structural configurations of the sensors. Apart from their bioresorbable nature, some of the biggest advantages of these sensors lie in their small size, their capability of performing precise measurements in delicate parts of the body, their consistency in the results, and the very low dependence of their responses on the change in temperature.

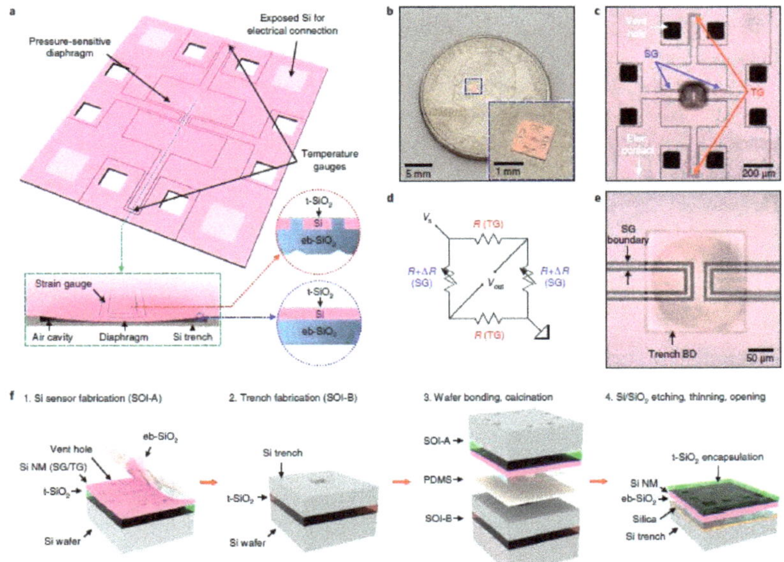

Figure 5. Schematic diagram of the fabrication process of the bioresorbable pressure sensors. The sensors were developed with monocrystalline silicon and silicon dioxide layers using thermal oxidation and electron-beam techniques. Two boundary lines were used to separate the strain gauges from the surrounding silicon. The system consisted of piezoresistive sensors, a voltage source and a meter to compensate for the variation of the temperature caused by the variation in the resistance with the pressure [95]. The image has been reproduced with permission from [95].

3. Current Challenges and Future Opportunities

Although a lot of work has been done in relation to silicon-based biomedical sensing, there are still some existing challenges that need attention in the current scenario. Starting from the fabrication of single-crystal silicon for forming wafers, it is an expensive process and requires highly-equipped refineries. In today's world, these refineries cost more than three billion dollars [96]. This is the reason why there is not much change in the cost of a single silicon chip. Second, in comparison to other substrates, silicon-based sensors, if produced in a lower quantity, would cost more per unit, unless ordered in bulk. Furthermore, toxic by-products are produced during the fabrication of silicon wafers [97]. This reduces their chances of being considered as useful for intrinsic biomedical applications. In comparison to the currently available organic conductive materials that have a high aspect ratio, silicon has a smaller sensing surface area, which eventually reduces its performance in terms of sensitivity and efficiency. The increase in the surface area would require an increase in the total size of the sensor, thus increasing the cost of the sensor per unit. Third, the thin-film wafers are very brittle, thus making it difficult to integrate them with signal-conditioning circuits in a more cost-efficient way. Another major disadvantage of the silicon-based sensors is the dependence of their responses on the change in temperature. Even though the sensors do perform well within the range of the temperature of the human body, these sensors possess a limitation when used for monitoring anatomical changes inside the human body. Additionally, due to the semiconducting nature of silicon, it is one of the least feasible options for forming the electrodes with the presence of highly conductive materials like gold, aluminium and others. One of the major factors that is limiting the use of silicon sensors for biomedical sensing in the current scenario is the limitation of its biocompatibility in comparison to the currently available carbon compounds. The biocompatibility should be increased via integrating or doping the silicon sensors to make them further suitable for implantable applications. Their response toward the analyte molecule also needs to be increased to decrease the interference of other elements during the experimental process. This is particularly necessary in the case of in-vivo applications, where the anatomical system of human beings and other animals is very complex. The inclusion of the presence of other bio-analyte, even in a very small amount on the sensing surface, can result in erroneous results. Other disadvantages of silicon-based sensors include their irregular behaviour and high signal-to-noise ratio (SNR) at a low frequency. This creates other problems at a higher frequency, such as a high input power and inconsistency in the data.

One of the remedial measures that can be considered to tackle the challenges mentioned above is the amalgamation of silicon with other materials to form both the substrates and electrodes of the sensors. For example, researchers have been making nanomaterials like silicon carbide on a large scale to form the electrodes for biomedical sensing [98]. The electrical and thermal characteristics of these nano-compounds can be affected by including certain elements like nitrogen and aluminium as impurities during the fabrication process. Similar to the conductive parts, silicon substrates would also be advantageous, especially for biomedical sensing. The flexibility of the substrates would allow for interfacing with certain intricate organs like the brain, in order to determine the signals from the soft tissues [99]. In terms of fabrication, doped-silicon electrodes can be formed on the prototypes by using soft lithographic techniques. Silicon can also be used as the master mould to form the designated electrode designs. This would be advantageous in terms of cost-operating conditions. Certain standard polymers like PDMS can be used as the stamp to form the final prototypes at room temperature. Furthermore, techniques like soft lithography can assist in quickly and efficiently obtaining 3D structures. The premise of using doped-silicon, along with silicon substrates, would also minimise the probability of affecting the characteristics of the final prototypes. This technique has a potential that would prove it to be better than the conventional photolithographic technique in terms of its resistance to the diffraction of light. Another primary advantage of using this technique is the possibility of the development of new 3D structures for biomedical applications that cannot be easily obtained by commonly used techniques. Researchers have been working [100,101] on this technique in conjugation with softer materials like organic polymers to obtain hybrid materials for healthcare applications. The sensors developed with this method have also performed with a high robustness, high efficiency, and enhanced mechanical and thermal characteristics [102]. Another solution for silicon-based sensors related to the high SNR at lower frequencies can be the use of biodegradable [103] and non-biodegradable [104] silicon nanowires for monitoring purposes. An alteration of the process of fabrication of silicon wafers can also be implemented to reduce the production of harmful by-products. This would also have an influence on their uses for biomedical sensing, as optimised fabrication processes would generate wafers with a better biocompatibility. One of the techniques for doing this is the fabrication of compound semiconducting wafers, which include materials such as Gallium Nitride, whose properties are similar to silicon. To reduce the SNR of silicon wafers, their sizes can be reduced to below 300 mm [105], which will not only allow an amplified signal for a high range of frequencies, but will also allow the reduction of the required power to drive the sensing systems. The application of the prototypes for biomedical sensing can be increased by making the silicon sensors multi-functional, operating on different genres. For example, an array of sensors can be formed to induce both electrochemical and strain-sensing purposes with the same system. This will reduce the total cost of the system while simultaneously improving their dynamic nature and portability. The range of parameters in biomedical sensing that are being monitored and diagnosed by the current sensors can be increased by including selectivity for specific applications. Sensors should be developed where the electrodes are intermixed with the template molecule for a particular analyte. In this way, the process of separately coating the sensing surface of the sensors can be eliminated.

The market surveys done on silicon-based sensors predict growth in their usage for biomedical sensing in the next few years. It has been estimated that the use of silicon would be extensively implemented to develop different kinds of medical devices, including pharmaceuticals, implants, and others [106]. The use of sensors in the form of microfluidic devices has been estimated to rise in fifteen years [107]. These silicon-based devices have been mostly used for pharmaceuticals, drug delivery, and in-vitro analyses. The market for silicon-based microfluidic devices was seen to be around 10.2 billion USD by 2017 and is predicted to rise in the future. The silicon sensors would also be used in the form of silicon carbide to make sensors for health care operations. It has been estimated that the rise in the use of silicon carbide materials would increase more than 1 billion USD by 2025, with a compound annual growth rate of 18.15% [108]. This increase in the cost of silicon sensors could indicate a rise in their subsequent uses in developing medical devices of a varied type. It is thus anticipated that there will be a prominent use of silicon-based sensing technology in the biomedical field for the ubiquitous monitoring of different acute and chronic diseases.

4. Conclusions

This paper explains the significance of the employment of silicon-based sensors for biomedical applications in the last two decades. Some of the important research in this area has been highlighted to define the fabrication and utilisation of different kinds of silicon-based prototypes for biomedical sensing purposes. The use of silicon to develop sensors has been advantageous in terms of their high abundance, low SNR, high thermal stability, resistance toward response changes based on ambiance conditions, high sensitivity and reliability, high repeatability in their response, low variation of response with time, low response time, and huge prosperity in terms of the future market. Apart from this, the electrical conductivity and mechanical flexibility of silicon have varied because of different methods, like the doping of elements to use them for varied purposes. Right from the fabrication of semiconducting substrates in the conventional MEMS technology, they have also been used to form nanowires in the prototype of the sensor. The above sensors have shown responses to in-vitro and in-vivo experiments for electrochemical analyte from animals and human beings. The use of silicon sensors for multifunctional applications is something that needs to be targeted in the future. This will not only minimise the cost of fabrication of the sensing systems but will also reduce the amount of generated electronic waste. The fundamental characteristics of silicon need to be further altered to increase their dynamicity in terms of the types of fabricated prototypes. The miniaturisation of sensors in the current world is one area where silicon-based sensors need to be fitted to utilise their advantages, as mentioned above. The conjugation of silicon with other conducting and semiconducting elements will assist in developing nanoparticle-based prototypes to further determine the minuscule intrinsic changes taking place in the body. The behaviour of these integrated sensors could, to a large extent, dictate their quality in the monitoring of biomedical applications.

Author Contributions: Conception and design, Y.X., A.N., N.A., S.K.; collection and assembly of data, Y.X., A.N., S.K., N.A., S.S.; manuscript writing, Y.X., A.N., S.K., N.A.; supervision, S.M. and Y.X; funding, Y.X, X.H., T.H.

Funding: This work was supported in part by International Scientific and Technological Cooperation Project of Dongguan (2016508102011), in part by Guangdong Provincial Key Platform and major Scientific Research Projects (2017GXJK174), in part by Science and Technology Innovation Special Fund Project of Guangdong Province (2018A050506086), in part by National Natural Science Foundation of China (61804028), in part by Dongguan core technology frontier project (2019622140003) and in part by Guangdong Natural Science Foundation (2018A030313492).

Acknowledgments: The authors are thankful to Anupriya Verma, professional English editor for her assistance in improving the quality of the manuscript in terms of its language.

Conflicts of Interest: The authors declare no conflict of interest.

References

1. Bhansali, S.; Vasudev, A. *MEMS for Biomedical Applications*; Elsevier: Amsterdam, The Netherlands, 2012.

2. Adhikari, K.K.; Qiang, T.; Wang, C.; Sung, H.K.; Wang, L.; Wu, Q. High-sensitivity radio frequency noncontact sensing and accurate quantification of uric acid in temperature-variant aqueous solutions. *Appl. Phys. Express* **2018**, *11*, 117001. [CrossRef]

3.   Sze, S.M. *Semiconductor Sensors*; Wiley: New York, NY, USA, 1994; Volume 55.

4.   Serene, M.; Babu, R.; Alex, Z.C. Sensitivity Analysis of Micro-Mass Optical MEMS Sensor for Biomedical IoT Devices. In *Internet of Things and Personalized Healthcare Systems*; Springer: Singapore, 2019.

5.   Qiang, T.; Wang, C.; Liu, M.Q.; Adhikari, K.K.; Liang, J.G.; Wang, L.; Li, Y.; Wu, Y.M.; Yang, G.H.; Meng, F.Y.; et al. High-Performance porous MIM-type capacitive humidity sensor realized via inductive coupled plasma and reactive-Ion etching. *Sens. Actuators B Chem.* **2018**, *258*, 704–714. [CrossRef]

6.   Kumari, S.K.; Mathana, J.M. Blood Sugar Level Indication Through Chewing and Swallowing from Acoustic MEMS Sensor and Deep Learning Algorithm for Diabetic Management. *J. Med. Syst.* **2019**, *43*, 1. [CrossRef] [PubMed]

7.   Gómez, V.; Soto Rodriguez, P.; Kumar, P.; Zaman, S.; Willander, M.; Nötzel, R. An InN/InGaN Quantum Dot Electrochemical Biosensor for Clinical Diagnosis. *Sensors* **2013**, *13*, 13917–13927.

8.   Rodriguez, P.E.S.; Gómez, V.J.; Kumar, P.; Willander, M.; Nötzel, R. Highly sensitive and fast anion selective InN quantum dot electrochemical sensors. *Appl. Phys. Express* **2013**, *6*, 115201.

9.   Alvi, N.H.; Soto Rodriguez, P.E.D.; Gómez, V.J.; Kumar, P.; Amin, G.; Nur, O.; Willander, M.; Nötzel, R. Highly Efficient Potentiometric Glucose Biosensor Based on Functionalized InN Quantum Dots. *Appl. Phys. Lett.* **2012**, *101*, 153110. [CrossRef]

10. Harraz, F.A. Porous silicon chemical sensors and biosensors: A review. *Sens. Actuators B Chem.* **2014**, *202*, 897–912. [CrossRef]

11. Santos, H.A. *Porous Silicon for Biomedical Applications*; Elsevier: Amsterdam, The Netherlands, 2014.

12. Silicon Wafer Production Process. Available online: https://www.sas-globalwafers.co.jp/eng/products/wafer/process.html (accessed on 27 June 2019).

13. Saliterman, S.S. *Fundamentals of BioMEMS and Medical Microdevices*; Wiley-Interscience: Bellingham, WA, USA, 2006.

14. Wang, Z.; Lee, S.; Koo, K.-I.; Kim, K. Nanowire-based sensors for biological and medical applications. *IEEE Trans. Nanobiosci.* **2016**, *15*, 186–199. [CrossRef]

15. Pramanik, C.; Saha, H. Low pressure piezoresistive sensors for medical electronics applications. *Mater. Manuf. Process.* **2006**, *21*, 233–238. [CrossRef]

16. Nag, A.; Zia, A.I.; Mukhopadhyay, S.; Kosel, J. Performance enhancement of electronic sensor through mask-less lithography. In Proceedings of the 9th International Conference on Sensing Technology (ICST), Auckland, New Zealand, 8–10 December 2015; pp. 374–379.

17. Herzer, N.; Hoeppener, S.; Schubert, U.S. Fabrication of patterned silane based self-assembled monolayers by photolithography and surface reactions on silicon-oxide substrates. *Chem. Commun.* **2010**, *46*, 5634–5652. [CrossRef]

18. Chen, H.; Xue, M.; Mei, Z.; Bambang Oetomo, S.; Chen, W. A review of wearable sensor systems for monitoring body movements of neonates. *Sensors* **2016**, *16*, 2134. [CrossRef] [PubMed]

19. Ng, J.; Sahakian, A.V.; Swiryn, S. *Sensing and Documentation of Body Position during Ambulatory ECG Monitoring*; IEEE: Piscataway, NJ, USA, 2000.

20. Iwasaki, W.; Nogami, H.; Takeuchi, S.; Furue, M.; Higurashi, E.; Sawada, R. Detection of site-specific blood flow variation in humans during running by a wearable laser Doppler flowmeter. *Sensors* **2015**, *15*, 25507–25519. [CrossRef] [PubMed]

21. Nomura, S.; Hanasaka, Y.; Katsuda, Y.; Hirota, R.; Ishiguro, T.; Takada, K.; Uryu, M.; Ogawa, H. Human pulse detection using multiple silicon microphones toward estimation of physical condition. *Inf. Technol. J.* **2012**, *11*, 476–479. [CrossRef]

22. Li, Y.Y.; Cunin, F.; Link, J.R.; Gao, T.; Betts, R.E.; Reiver, S.H.; Chin, V.; Bhatia, S.N.; Sailor, M.J. Polymer replicas of photonic porous silicon for sensing and drug delivery applications. *Science* **2003**, *299*, 2045–2047. [CrossRef] [PubMed]

23. Mariani, S.; Pino, L.; Strambini, L.M.; Tedeschi, L.; Barillaro, G. 10000-fold improvement in protein detection using nanostructured porous silicon interferometric aptasensors. *ACS Sens.* **2016**, *1*, 1471–1479. [CrossRef]

24. Washburn, A.L.; Shia, W.W.; Lenkeit, K.A.; Lee, S.-H.; Bailey, R.C. Multiplexed cancer biomarker detection using chip-integrated silicon photonic sensor arrays. *Analyst* **2016**, *141*, 5358–5365. [CrossRef]

25. Fritz, J.; Cooper, E.B.; Gaudet, S.; Sorger, P.K.; Manalis, S.R. Electronic detection of DNA by its intrinsic molecular charge. *Proc. Natl. Acad. Sci. USA* **2002**, *99*, 14142–14146. [CrossRef]

26. Rajagopalan, J.; Saif, M.T.A. MEMS sensors and microsystems for cell mechanobiology. *J. Micromech. Microeng.* **2011**, *21*, 054002. [CrossRef]

27. Ciuti, G.; Ricotti, L.; Menciassi, A.; Dario, P. MEMS sensor technologies for human centred applications in healthcare, physical activities, safety and environmental sensing: A review on research activities in Italy. *Sensors* **2015**, *15*, 6441–6468. [CrossRef]

28. Sharma, H.; Alvi, P.; Dalela, S.; Akhtar, J. Role of MEMS in biomedical application: A review. *Sens. Transducers* **2010**, *115*, 1.

29. Jivani, R.R.; Lakhtaria, G.J.; Patadiya, D.D.; Patel, L.D.; Jivani, N.P.; Jhala, B.P. Biomedical microelectromechanical systems (BioMEMS): Revolution in drug delivery and analytical techniques. *Saudi Pharm. J.* **2016**, *24*, 1–20. [CrossRef] [PubMed]

30. Grayson, A.C.R.; Shawgo, R.S.; Johnson, A.M.; Flynn, N.T.; Li, Y.; Cima, M.J.; Langer, R. A BioMEMS review: MEMS technology for physiologically integrated devices. *Proc. IEEE* **2004**, *92*, 6–21. [CrossRef]

31. Judy, J.W. Biomedical applications of MEMS. In Proceedings of the Measurement and Science Technology Conference, Anaheim, CA, USA, 31 January–1 February 1991; pp. 403–414.

32. Arshak, A.; Arshak, K.; Lyons, G.; Waldron, D.; Morris, D.; Korostynska, O.; Jafer, E. Review of the potential of a wireless MEMS microsystem for biomedical applications. *Sens. Rev.* **2005**, *25*, 277–286. [CrossRef]

33. Wang, W.; Soper, S.A. *Bio-MEMS: Technologies and Applications*; CRC Press: Boca Raton, FL, USA, 2006.

34. Dheringe, N.; Rahane, S. Recent advances in mems sensor technology biomedical mechanical thermo-fluid & electromagnetic sensors. *Int. J. Electron. Commun. Instrum. Eng. Res. Dev.* **2013**, *3*, 73–90.

35. He, B.; Morrow, T.J.; Keating, C.D. Nanowire sensors for multiplexed detection of biomolecules. *Curr. Opin. Chem. Biol.* **2008**, *12*, 522–528. [CrossRef] [PubMed]

36. Schander, A.; Stemmann, H.; Kreiter, A.; Lang, W. Silicon-Based Microfabrication of Free-Floating Neural Probes and Insertion Tool for Chronic Applications. *Micromachines* **2018**, *9*, 131. [CrossRef] [PubMed]

37. Boric-Lubecke, O.; Droitcour, A.D.; Lubecke, V.M.; Lin, J.; Kovacs, G.T. Wireless IC Doppler radars for sensing of heart and respiration activity. In Proceedings of the 6th International Conference on Telecommunications in Modern Satellite, Cable and Broadcasting Service, Niš, Yugoslavia, 1–3 October 2003; pp. 337–344.

38. Weisfield, R.L.; Hartney, M.A.; Street, R.A.; Apte, R.B. New amorphous-silicon image sensor for x-ray diagnostic medical imaging applications. In Proceedings of the Medical Imaging: Physics of Medical Imaging, San Diego, CA, USA, 22–24 February 1998; pp. 444–453.

39. Kovacs, G.T.; Storment, C.W.; Halks-Miller, M.; Belczynski, C.R.; Santina, C.D.; Lewis, E.R.; Maluf, N.I. Silicon-substrate microelectrode arrays for parallel recording of neural activity in peripheral and cranial nerves. *IEEE Trans. Biomed. Eng.* **1994**, *41*, 567–577. [CrossRef] [PubMed]

40. Borkholder, D. Cell based Biosensors Using Microelectrodes. Ph.D. Thesis, Stanford University, Stanford, CA, USA, 1998.

41. Ksendzov, A.; Lin, Y. Integrated optics ring-resonator sensors for protein detection. *Opt. Lett.* **2005**, *30*, 3344–3346. [CrossRef]

42. Gary, P.A. *Modeling and Optimization of a Silicon Photosensor for a Reading Aid*; Stanford University CA Stanford Electronics Labs: Stanford, CA, USA, 1967.

43. Hill, D. Progress in medical instrumentation over the past fifty years. *J. Phys. E Sci. Instrum.* **1968**, *1*, 697. [CrossRef]

44. Wise, K.D. Integrated sensors, MEMS, and microsystems: Reflections on a fantastic voyage. *Sens. Actuators A Phys.* **2007**, *136*, 39–50. [CrossRef]

45. Kotzar, G.; Freas, M.; Abel, P.; Fleischman, A.; Roy, S.; Zorman, C.; Moran, J.M.; Melzak, J. Evaluation of MEMS materials of construction for implantable medical devices. *Biomaterials* **2002**, *23*, 2737–2750. [CrossRef]

46. Durini, D. *High Performance Silicon Imaging: Fundamentals and Applications of CMOS and CCD Sensors*; Elsevier: Amsterdam, The Netherlands, 2014.

47. Smith, C.S. Piezoresistance effect in germanium and silicon. *Phys. Rev.* **1954**, *94*, 42. [CrossRef]

48. Tufte, O.; Chapman, P.; Long, D. Silicon diffused-element piezoresistive diaphragms. *J. Appl. Phys.* **1962**, *33*, 3322–3327. [CrossRef]

49. Afsarimanesh, N.; Mukhopadhyay, S.C.; Kruger, M. Molecularly imprinted polymer-based electrochemical biosensor for bone loss detection. *IEEE Trans. Biomed. Eng.* **2018**, *65*, 1264–1271. [CrossRef] [PubMed]

50. Nag, A.; Zia, A.I.; Li, X.; Mukhopadhyay, S.C.; Kosel, J. Novel Sensing Approach for LPG Leakage Detection: Part I—Operating Mechanism and Preliminary Results. *IEEE Sens. J.* **2016**, *16*, 996–1003. [CrossRef]

51. Nag, A.; Zia, A.I.; Li, X.; Mukhopadhyay, S.C.; Kosel, J. Novel Sensing Approach for LPG Leakage Detection—Part II: Effects of Particle Size, Composition, and Coating Layer Thickness. *IEEE Sens. J.* **2016**, *16*, 1088–1094. [CrossRef]

52. Van Putten, M.; Van Putten, M.; Van Putten, A.; Pompe, J.; Bruining, H. A silicon bidirectional flow sensor for measuring respiratory flow. *IEEE Trans. Biomed. Eng.* **1997**, *44*, 205–208. [CrossRef]

53. Zhou, X.; Hu, J.; Li, C.; Ma, D.; Lee, C.; Lee, S. Silicon nanowires as chemical sensors. *Chem. Phys. Lett.* **2003**, *369*, 220–224. [CrossRef]

54. Otte, N. The silicon photomultiplier-a new device for high energy physics, astroparticle physics, industrial and medical applications. In Proceedings of the IX International Symposium on Detectors for Particle, Astroparticle and Synchrotron Radiation Experiments, Stanford, CA, USA, 3–6 April 2006; pp. 1–9.

55. Laschi, S.; Mascini, M. Planar electrochemical sensors for biomedical applications. *Med. Eng. Phys.* **2006**, *28*, 934–943. [CrossRef]

56. Mamishev, A.V.; Sundara-Rajan, K.; Yang, F.; Du, Y.; Zahn, M. Interdigital sensors and transducers. *Proc. IEEE* **2004**, *92*, 808–845. [CrossRef]

57. Diamond, G.G.; Hutchins, D.A. A new capacitive imaging technique for NDT. In Proceedings of the European Conference NDT, Berlin, Germany, 25–29 September 2006; p. 229.

58. Osoinach, B. Proximity Capacitive Sensor Technology for Touch Sensing Applications. Available online: cache.freescale.com/files/sensors/doc/white_paper/PROXIMITYWP.pdf (accessed on 27 June 2019).

59. Anandan, N.; George, B. A wide-range capacitive sensor for linear and angular displacement measurement. *IEEE Trans. Ind. Electron.* **2017**, *64*, 5728–5737. [CrossRef]

60. Afsarimanesh, N.; Mukhopadhyay, S.C.; Kruger, M. Planar Interdigital Sensors and Electrochemical Impedance Spectroscopy. In *Electrochemical Biosensor: Point-of-Care for Early Detection of Bone Loss*; Springer: Berlin, Germany, 2019; pp. 33–44.

61. Cheng, B. Security Imaging Devices with Planar Capacitance Sensor Arrays. Ph.D. Thesis, University of Manchester, Manchester, UK, 2008.

62. Frounchi, J.; Dehkhoda, F. High-speed capacitance scanner. In Proceedings of the 3rd World Congress on Industrial Process Tomography, Banff, AB, Canada, 2–5 September 2003; pp. 846–852.

63. Hu, X.; Yang, W. Planar capacitive sensors–Designs and applications. *Sens. Rev.* **2010**, *30*, 24–39. [CrossRef]

64. Musayev, J.; Adlgüzel, Y.; Külah, H.; Eminoğlu, S.; Akln, T. Label-free DNA detection using a charge sensitive CMOS microarray sensor chip. *IEEE Sens. J.* **2014**, *14*, 1608–1616. [CrossRef]

65. Afsarimanesh, N.; Mukhopadhyay, S.C.; Kruger, M. Performance Assessment of Interdigital Sensor for Varied Coating Thicknesses to Detect CTX-I. *IEEE Sens. J.* **2018**, *18*, 3924–3931. [CrossRef]

66. Afsarimanesh, N.; Alahi, M.E.E.; Mukhopadhyay, S.C.; Kruger, M. Development of IoT-Based Impedometric Biosensor for Point-of-Care Monitoring of Bone Loss. *IEEE J. Emerg. Sel. Top. Circuits Syst.* **2018**, *8*, 211–220. [CrossRef]

67. Fernandez, R.E.; Stolyarova, S.; Chadha, A.; Bhattacharya, E.; Nemirovsky, Y. MEMS composite porous silicon/polysilicon cantilever sensor for enhanced triglycerides biosensing. *IEEE Sens. J.* **2009**, *9*, 1660–1666. [CrossRef]

68. Abdelghani, L.; Nasr-Eddine, M.; Azouza, M.; Abdellah, B.; Moadh, K. Modeling of silicon MEMS capacitive pressure sensor for biomédical applications. In Proceedings of the 9th International Design and Test Symposium (IDT), Algiers, Algeria, 16–18 December 2014; pp. 263–266.

69. Park, J.; Kim, J.-K.; Kim, D.-S.; Shanmugasundaram, A.; Park, S.A.; Kang, S.; Kim, S.-H.; Jeong, M.H.; Lee, D.-W. Wireless pressure sensor integrated with a 3D printed polymer stent for smart health monitoring. *Sens. Actuators B Chem.* **2019**, *280*, 201–209. [CrossRef]

70. Tohyama, O.; Kohashi, M.; Sugihara, M.; Itoh, H. A fiber-optic pressure microsensor for biomedical applications. *Sens. Actuators A Phys.* **1998**, *66*, 150–154. [CrossRef]

71. Reinhoudt, D. Durable chemical sensors based on field-effect transistors. *Sens. Actuators B Chem.* **1995**, *24*, 197–200. [CrossRef]

72. Chudy, M.; Wroblewski, W.; Dybko, A.; Brzozka, Z. Multi-ion analysis based on versatile sensor head. *Sens. Actuators B Chem.* **2001**, *78*, 320–325. [CrossRef]

73. Sergeyeva, T.; Soldatkin, A.; Rachkov, A.; Tereschenko, M.; Piletsky, S.; Elskaya, A. β-Lactamase label-based potentiometric biosensor for α-2 interferon detection. *Anal. Chim. Acta* **1999**, *390*, 73–81. [CrossRef]

74. Selvanayagam, Z.E.; Neuzil, P.; Gopalakrishnakone, P.; Sridhar, U.; Singh, M.; Ho, L. An ISFET-based immunosensor for the detection of β-Bungarotoxin. *Biosens. Bioelectron.* **2002**, *17*, 821–826. [CrossRef]

75. Errachid, A.; Ivorra, A.; Aguilo, J.; Villa, R.; Zine, N.; Bausells, J. New technology for multi-sensor silicon needles for biomedical applications. *Sens. Actuators B Chem.* **2001**, *78*, 279–284. [CrossRef]

76. Young, D.; Cong, P. Wireless Implantable Sensors: From Lab to Technology Breakthrough Ambitions. *Sens. Actuators A Phys.* **2019**, *294*, 81–90. [CrossRef]

77. Kobayashi, T.; Makimoto, N.; Oikawa, T.; Wada, A.; Funakubo, H.; Maeda, R. Linear actuation piezoelectric microcantilever using tetragonal composition PZT thin films. In Proceedings of the IEEE 26th International Conference on Micro Electro Mechanical Systems (MEMS), Taipei, Taiwan, 20–24 January 2013; pp. 413–416.

78. Kobayashi, T.; Oyama, S.; Okada, H.; Makimoto, N.; Tanaka, K.; Itoh, T.; Maeda, R. An electrostatic field sensor driven by self-excited vibration of sensor/actuator integrated piezoelectric micro cantilever. In Proceedings of the IEEE 25th International Conference on Micro Electro Mechanical Systems (MEMS), Paris, France, 29 January–2 February 2012; pp. 527–530.

79. Feng, W.; Hettiarachchi, R.; Sato, S.; Kakushima, K.; Niwa, M.; Iwai, H.; Yamada, K.; Ohmori, K. Advantages of Silicon Nanowire MOSFETs over Planar Ones Investigated from the Viewpoints of Static and Noise Properties. In Proceedings of the International Conference on Solid State Devices and Materials, Nagoya, Japan, 2–5 September 2011.

80. Zhu, H. Semiconductor Nanowire MOSFETs and Applications. *Nanowires New Insights* **2017**, *101*. [CrossRef]

81. Chen, K.-I.; Li, B.-R.; Chen, Y.-T. Silicon nanowire field-effect transistor-based biosensors for biomedical diagnosis and cellular recording investigation. *Nano Today* **2011**, *6*, 131–154. [CrossRef]

82. Abdolahad, M.; Taghinejad, H.; Saeidi, A.; Taghinejad, M.; Janmaleki, M.; Mohajerzadeh, S. Cell membrane electrical charge investigations by silicon nanowires incorporated field effect transistor (SiNWFET) suitable in cancer research. *RSC Adv.* **2014**, *4*, 7425–7431. [CrossRef]

83. Zhang, G.-J.; Zhang, G.; Chua, J.H.; Chee, R.-E.; Wong, E.H.; Agarwal, A.; Buddharaju, K.D.; Singh, N.; Gao, Z.; Balasubramanian, N. DNA sensing by silicon nanowire: Charge layer distance dependence. *Nano Lett.* **2008**, *8*, 1066–1070. [CrossRef]

84. Jamaa, M.H.B.; Carrara, S.; Georgiou, J.; Archontas, N.; De Micheli, G. Fabrication of memristors with poly-crystalline silicon nanowires. In Proceedings of the 9th IEEE Conference on Nanotechnology (IEEE-NANO), Genoa, Italy, 26–30 July 2009; pp. 152–154.

85. Tzouvadaki, I.; Parrozzani, C.; Gallotta, A.; De Micheli, G.; Carrara, S. Memristive biosensors for PSA-IgM detection. *BioNanoScience* **2015**, *5*, 189–195. [CrossRef]

86. Tzouvadaki, I.; Puppo, F.; Doucey, M.-A.; De Micheli, G.; Carrara, S. Computational study on the electrical behavior of silicon nanowire memristive biosensors. *IEEE Sens. J.* **2015**, *15*, 6208–6217. [CrossRef]

87. Beccai, L.; Roccella, S.; Arena, A.; Valvo, F.; Valdastri, P.; Menciassi, A.; Carrozza, M.C.; Dario, P. Design and fabrication of a hybrid silicon three-axial force sensor for biomechanical applications. *Sens. Actuators A Phys.* **2005**, *120*, 370–382. [CrossRef]

88. Manikandan, N.; Muruganand, S.; Divagar, M.; Viswanathan, C. Design and fabrication of MEMS based intracranial pressure sensor for neurons study. *Vacuum* **2019**, *163*, 204–209. [CrossRef]

89. Pramanik, C.; Saha, H.; Gangopadhyay, U. Design optimization of a high performance silicon MEMS piezoresistive pressure sensor for biomedical applications. *J. Micromech. Microeng.* **2006**, *16*, 2060. [CrossRef]

90. Wu, N.; Tian, Y.; Zou, X.; Zhai, Y.; Barringhaus, K.; Wang, X. A miniature fiber optic blood pressure sensor and its application in in vivo blood pressure measurements of a swine model. *Sens. Actuators B Chem.* **2013**, *181*, 172–178. [CrossRef]

91. Satake, D.; Ebi, H.; Oku, N.; Matsuda, K.; Takao, H.; Ashiki, M.; Ishida, M. A sensor for blood cell counter using MEMS technology. *Sens. Actuators B Chem.* **2002**, *83*, 77–81. [CrossRef]

92. Kang, S.-K.; Murphy, R.K.; Hwang, S.-W.; Lee, S.M.; Harburg, D.V.; Krueger, N.A.; Shin, J.; Gamble, P.; Cheng, H.; Yu, S. Bioresorbable silicon electronic sensors for the brain. *Nature* **2016**, *530*, 71–76. [CrossRef] [PubMed]

93. Fang, H.; Zhao, J.; Yu, K.J.; Song, E.; Farimani, A.B.; Chiang, C.-H.; Jin, X.; Xue, Y.; Xu, D.; Du, W. Ultrathin, transferred layers of thermally grown silicon dioxide as biofluid barriers for biointegrated flexible electronic systems. *Proc. Natl. Acad. Sci. USA* **2016**, *113*, 11682–11687. [CrossRef]

94. Gabriel, G.; Erill, I.; Caro, J.; Gómez, R.; Riera, D.; Villa, R.; Godignon, P. Manufacturing and full characterization of silicon carbide-based multi-sensor micro-probes for biomedical applications. *Microelectron. J.* **2007**, *38*, 406–415. [CrossRef]

95. Shin, J.; Yan, Y.; Bai, W.; Xue, Y.; Gamble, P.; Tian, L.; Kandela, I.; Haney, C.R.; Spees, W.; Lee, Y. Bioresorbable pressure sensors protected with thermally grown silicon dioxide for the monitoring of chronic diseases and healing processes. *Nat. Biomed. Eng.* **2019**, *3*, 37. [CrossRef]

96. Disadvantages of Silicon Sensors. Available online: https://www.printedelectronicsworld.com/articles/52/problems-with-silicon-chips (accessed on 27 June 2019).

97. Advantages and Disadvantages of Silicon. Available online: http://www.rfwireless-world.com/Terminology/Advantages-and-Disadvantages-of-Silicon.html (accessed on 27 June 2019).

98. Han, T.; Kundu, S.; Nag, A.; Xu, Y. 3D Printed Sensors for Biomedical Applications: A Review. *Sensors* **2019**, *19*, 1706. [CrossRef]

99. Gupta, S.; Navaraj, W.T.; Lorenzelli, L.; Dahiya, R. Ultra-thin chips for high-performance flexible electronics. *NPJ Flex. Electron.* **2018**, *2*, 8. [CrossRef]

100. Herbert, R.; Kim, J.-H.; Kim, Y.; Lee, H.; Yeo, W.-H. Soft material-enabled, flexible hybrid electronics for medicine, healthcare, and human-machine interfaces. *Materials* **2018**, *11*, 187. [CrossRef] [PubMed]

101. Khan, S.M.; Gumus, A.; Nassar, J.M.; Hussain, M.M. CMOS enabled microfluidic systems for healthcare based applications. *Adv. Mater.* **2018**, *30*, 1705759. [CrossRef] [PubMed]

102. Kang, S.-W. Application of Soft Lithography for Nano Functional Devices. *Lithography* **2010**, 403.

103. Chiappini, C.; Liu, X.; Fakhoury, J.R.; Ferrari, M. Biodegradable porous silicon barcode nanowires with defined geometry. *Adv. Funct. Mater.* **2010**, *20*, 2231–2239. [CrossRef] [PubMed]

104. Rajan, N.K.; Routenberg, D.A.; Reed, M.A. Optimal signal-to-noise ratio for silicon nanowire biochemical sensors. *Appl. Phys. Lett.* **2011**, *98*, 264107. [CrossRef] [PubMed]

105. Changing Wafer Size and the Move to 300 mm. Available online: http://smithsonianchips.si.edu/ice/cd/CEICM/SECTION7.pdf (accessed on 27 June 2019).

106. Medical Flexible Packaging Market Analysis by Material. Available online: https://www.grandviewresearch.com/industry-analysis/medical-flexible-packaging-market (accessed on 27 June 2019).

107. Global Silicon Carbide Market. Available online: https://flairinsights.com/report-details/global-silicon-carbide-market/14663 (accessed on 27 June 2019).

108. Silicon Carbide Power Semiconductors Market Worth. Available online: https://www.prnewswire.com/news-releases/silicon-carbide-power-semiconductors-market-worth-1-109-mn-by-2025-cagr-18-1-allied-market-research-881380442.html (accessed on 27 June 2019).

Article

A Sub-mW 18-MHz MEMS Oscillator Based on a 98-dBΩ Adjustable Bandwidth Transimpedance Amplifier and a Lamé-Mode Resonator

Anoir Bouchami, Mohannad Y. Elsayed and Frederic Nabki *

Department of Electrical Engineering, École de technologie supérieure, Montréal, QC H3C 1K3, Canada;
bouchami.anoir@courrier.uqam.ca (A.B.); mohannad.elsayed@mail.mcgill.ca (M.Y.E.)
* Correspondence: frederic.nabki@etsmtl.ca

Received: 27 April 2019; Accepted: 11 June 2019; Published: 13 June 2019

Abstract: This paper presents a microelectromechanical system (MEMS)-based oscillator based on a Lamé-mode capacitive micromachined resonator and a fully differential high-gain transimpedance amplifier (TIA). The proposed TIA is designed using TSMC 65 nm CMOS technology and consumes only 0.9 mA from a 1-V supply. The measured mid-band transimpedance gain is 98 dBΩ and the TIA features an adjustable bandwidth with a maximum bandwidth of 142 MHz for a parasitic capacitance C_P of 4 pF. The measured input-referred current noise of the TIA at mid-band is below $15\,\mathrm{pA}/\sqrt{\mathrm{Hz}}$. The TIA is connected to a Lamé-mode resonator, and the oscillator performance in terms of phase noise and frequency stability is presented. The measured phase noise under vacuum is $-120\,\mathrm{dBc/Hz}$ at a 1-kHz offset, while the phase noise floor reaches $-127\,\mathrm{dBc/Hz}$. The measured short-term stability of the MEMS-based oscillator is ± 0.25 ppm.

Keywords: MEMS-based oscillator; transimpedance amplifier; electrostatic actuation; phase noise; Lamé-mode MEMS resonator; input-referred noise; quality factor; 1-dB compression point

1. Introduction

Oscillators are of great interest because of their ubiquitous use in timing applications and in modern wireless communication devices. They are indispensable to ensure proper synchronization in almost any system. Micromachined resonators, a subset of microelectromechanical systems (MEMS), are receiving continuously increasing interest due to their small sizes as well as their potential for integration with other integrated devices and circuits on the same chip which makes them excellent candidates for replacing crystal-based resonators in timing applications. This is especially important for handheld and wearable electronic applications where weight, size, and cost are critical parameters. However, resonator performance, mainly operation frequency (f_0), quality factor (Q), resonator parasitics and motional resistance, set stringent requirements for the oscillation sustaining circuitry (e.g., gain and power consumption) in order to achieve a high-performance MEMS oscillator.

Resonators can be classified based on their vibration modes as either flexural or bulk-mode devices. Bulk-mode devices typically exhibit high stiffness, and are consequently less prone to thermoelastic damping, compared to flexural devices, allowing them to achieve large quality factors (>10,000), even at atmospheric pressure [1–16]. In [2–7], bulk-mode resonators including Lamé-mode and wine glass mode devices with quality factors in the 10^6 range were presented based on capacitive actuation. Such high quality factors are achievable as a result of the superior structural material, single crystalline silicon, which is patterned into a pure bulk resonating structure, without the need for release holes, or additional layers on top, which is one of the limitations for the quality factors achieved by piezoelectrically actuated bulk-mode devices (e.g., [8–10]). On the other hand, capacitive bulk-mode devices typically exhibit lower transduction efficiencies compared to piezoelectric devices, which translate to higher losses and motional resistances. This can be accounted for by either enhancing the transduction, e.g., sub-micron gaps realized by complex fabrication processes [11–13], high voltages [1–7,14], added transducer combs [15,16], movable electrodes for gap closing [17,18], or increasing the gain for the transimpedance amplifier (TIA) in order to sustain oscillation. Several transimpedance topologies have been reported in the literature for MEMS-based oscillator applications [19–26]. Designs proposed in [19–21] use an automatic gain control circuit to regulate the oscillation amplitude and reduce the resonator mechanical non-linearity effect. Furthermore, while differential TIAs allow for better performance, the power consumption of fully differential transimpedance amplifier designs in [19,22] are higher than single-ended TIAs in [20,21,23–25].

Typically, a MEMS oscillator is realized by connecting a TIA with the resonator in a positive feedback loop to sustain a steady-state oscillation by converting the resonator output current to an output voltage signal and providing sufficient gain and suitable phase. The quality of the output oscillation is usually determined by the quality factor of the resonator and by the electrical noise of the TIA. It is necessary for the TIA to have high transimpedance gain due to the insertion loss of the resonator caused by its motional resistance. A sufficiently wide bandwidth is also required to ensure that the oscillator phase shift is around $0°$, when the MEMS-based oscillator operates in series resonance. Thus, Barkhausen conditions will be fulfilled [27]. Furthermore, low input and output impedances are required to minimize the resonator Q-factor loading.

Electrostatic MEMS resonators typically exhibit relatively high loss which necessitates the use of a high-gain transimpedance amplifier in the feedback path to compensate for the losses. While an inverting amplifier can be used (i.e., Pierce oscillator structure), if the circuit bandwidth can be made sufficiently large, the oscillation frequency can be accurately generated with a non-inverting amplifier, without relying on external capacitors and their values in order to ensure the $180°$ additional phase shift required [28]. This also yields a more compact system, not requiring these off-chip capacitors.

This work proposes a MEMS oscillator that is based on the Lamé-mode resonator presented in [1]. The oscillator includes a fully differential high-gain TIA that is tailored to the resonator. Accordingly, the oscillator achieves very competitive performance in terms of power consumption and phase noise. The paper begins with a summary description of the Lamé-mode resonator followed by detailed descriptions of each of the TIA's building blocks. Measurement results are then presented and discussed, and are followed by a conclusion.

2. Lamé-Mode MEMS Resonator Overview

A brief description of the Lamé-mode capacitive (i.e., electrostatic) MEMS resonator presented in [1] is given in this section. Figure 1 illustrates exploded and assembled 3D renditions of the resonator structure. Structures were fabricated in a commercial silicon-on-insulator (SOI) technology, MicraGEM-Si, where they are realized through processing and wafer bonding of two SOI wafers (i.e., the top wafer and the bottom wafer). The top wafer has its handle layer removed after bonding to the bottom wafer such that the resonator is mainly composed of a single crystalline silicon central square suspended structure acting as the Lamé bulk-mode resonator. This suspended square is 30 μm thick, has a 230 μm side length, and is formed in the device layer of the top SOI wafer. The resonator square structure is anchored to the substrate through four suspension beams placed at the corner nodal points of the resonance mode. Pads for an electrical connection to the central square are present at the end of each suspension beam. This allows the connection of the DC polarization voltage required for the electrostatic actuation of the device. These support beams are patterned in the device layer of the top SOI wafer. The central structure is surrounded by four electrodes used for capacitive actuation and sensing of the structure. The electrodes are formed in the device layer of the top SOI wafer and are separated from the central square by a 2 μm capacitive transduction gap, which is the minimum spacing allowed by the technology. The device layer of the bottom SOI wafer is patterned to form the electrode anchors and the anchors at the end of the suspension beams. SEM micrographs of the resonator are shown in Figure 2. FEM simulations as well as theoretical calculations predict a resonance frequency of 17.9 MHz.

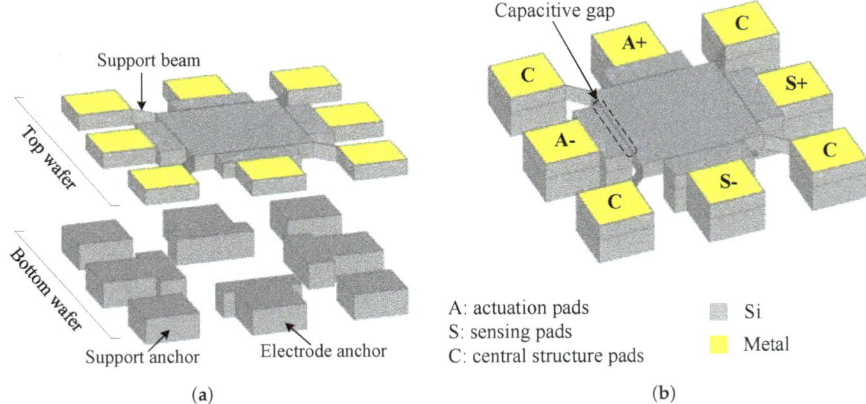

Figure 1. Simplified diagram of the (**a**) exploded and (**b**) assembled views of the Lamé-mode MEMS resonator with corner supports [1].

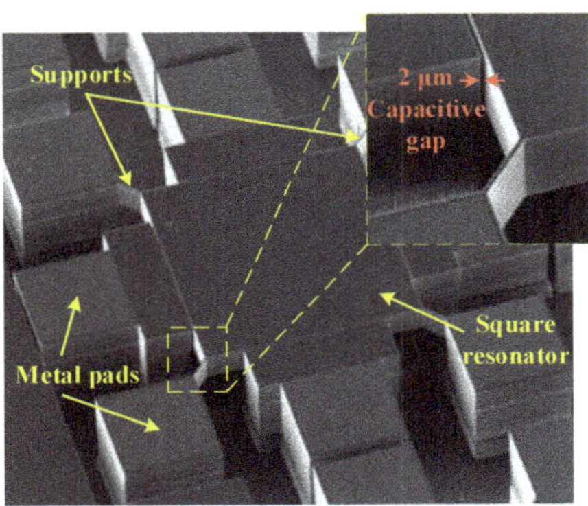

Figure 2. SEM micrograph of the Lamé-mode MEMS resonator with corner supports [1].

3. Transimpedance Amplifier Circuit Design

The TIA circuit shown in Figure 3 is composed of a three fully differential stages: (i) an input stage followed by (ii) a variable gain amplifier (VGA) controlled by an automatic gain control circuit (AGC), and (iii) an output stage implemented with a super source follower (SSF). The complete schematic circuit is shown in Figure 4 in which the biasing and common-mode feedback (CMFB) circuits are not shown. The sustaining amplifier provides low input impedance (R_{in}) and low output impedance (R_{out}) so as to compensate for large interconnect parasitic capacitance ($C_P = 4$ pF) and push the dominant pole far beyond the oscillation frequency. This translates into a high-gain-bandwidth (GBW) product [29].

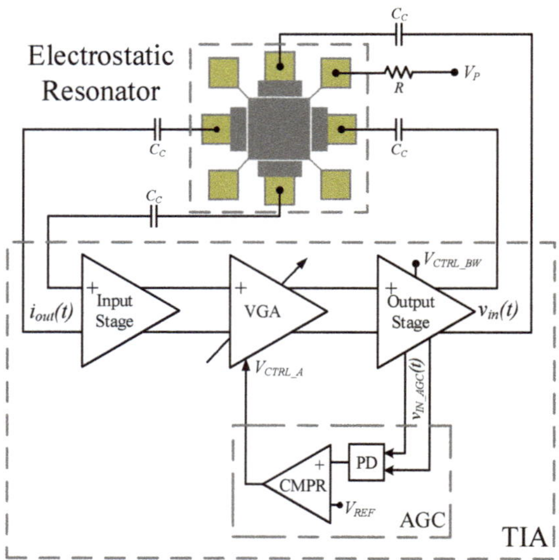

Figure 3. MEMS-based oscillator functional diagram.

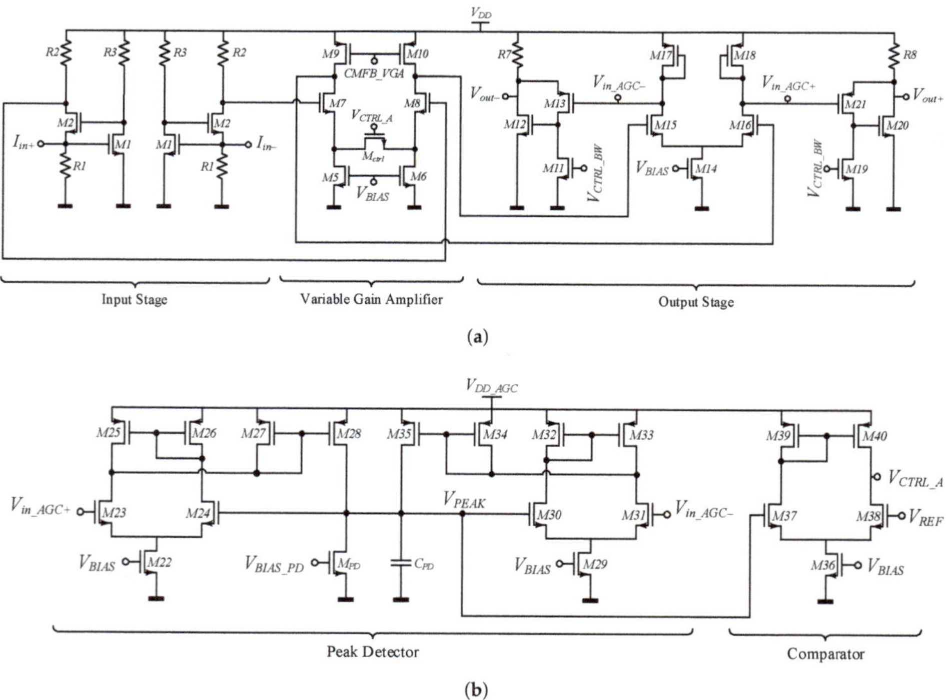

Figure 4. Circuit schematic of: (**a**) the proposed fully differential TIA design, and (**b**) the AGC.

The gain of the TIA needs to be high enough to compensate for the motional resistance of the MEMS resonator and sustain the oscillation. The regulated cascade (RGC) topology [30] was chosen as the input stage to achieve a reasonable trade-off between gain, bandwidth, and power consumption. The input impedance of the RGC input stage is given by

$$R_{in} = \frac{1}{g_{m2}\left(1 + R_3\, g_{m1}\right)},\tag{1}$$

where g_{m1} and g_{m2} are the transconductances of transistors $M1$ and $M2$, respectively. Thus, a smaller input impedance can be attained by increasing the voltage gain of the local feedback stage given by $(1 + R_3\, g_{m1})$. The gain of the input stage is given by

$$Z_T(s) \cong \frac{R_2\left(1 + s\dfrac{R_3 C_1}{1 + R_3 g_{m1}}\right)}{(1 + sR_1 C_{in})\,(1 + sR_3 C_1)\left(1 + sR_2 C_{gd2}\right)},\tag{2}$$

where C_{in}, C_1, and C_{gd2} are the total input capacitances of the input stage, the equivalent capacitance between the drain of $M1$ and gate of $M2$, and the gate-drain capacitance of transistors $M2$, respectively. To achieve a higher gain, R_2 should be increased, although it cannot be arbitrarily enlarged because of design constraints. It can be seen from Equation (2) that the 3-dB bandwidth of the input stage is limited by the dominant pole appearing at the drain of transistor $M1$ and is given by

$$f_{-3dB} = \frac{1}{2\pi R_1 C_{in}} \cong \frac{1}{2\pi R_1 \times \left(C_{gs1} + C_{gd1} R_3 g_{m1}\right)},\tag{3}$$

where C_{gd1} and C_{gs1} are the gate-drain capacitance and the gate-source capacitance of transistor $M1$, respectively. The local feedback of the input stage generates a zero at a frequency of

$$f_z \cong \frac{g_{m1}}{2\pi C_1} \cong \frac{g_{m1}}{2\pi\left[C_{gd1} + C_{gd2}\left(1 + \dfrac{R_2}{R_1}\right)\right]}.\tag{4}$$

To maintain the zero far away from the dominant pole [31], the gate-drain capacitance of transistor $M2$ should be reduced by decreasing its width. In this fashion, the RGC input impedance in Equation (1) will not be dramatically affected since g_{m2} will not decrease considerably as it is proportional to $\sqrt{(W/L)_2}$, while its gate capacitance is linearly proportional to $(WL)_2$. This can be compensated by increasing R_3 as the input impedance is inversely proportional to $(1 + R_3\, g_{m1})$, as shown in Equation (1).

The input-referred current noise is a key performance parameter to be considered when designing the proposed TIA. It can be used to provide a representative comparison between different circuit topologies. Since the noise is mostly contributed by the input stage, the noise of the other stages can be neglected. Therefore, a noise analysis is carried-out using the equivalent circuit shown in Figure 5, and is based on the analysis method proposed in [31], where shot noise and flicker noise are neglected. Assuming that all the noise sources are uncorrelated, the input-referred current noise of the input stage can be shown to be given by

$$\overline{i_{n,in}^2} = \frac{4kT}{R_1} + \frac{\omega^2(C_1 + C_2)^2}{g_{m2}^2}\left(\gamma g_{d0,2} + \frac{1}{R_2}\right) + \frac{4kT\left(\dfrac{1}{R_1^2} + \omega^2 C_{in}^2\right)}{\left(g_{m1} + \dfrac{1}{R_3}\right)^2}\left(\gamma g_{d0,1} + \frac{1}{R_3}\right),\tag{5}$$

where γ is the noise coefficient [32,33], k is Boltzmann's constant, T is the absolute temperature, and $g_{d0,1}$ and $g_{d0,2}$ are the zero-bias drain conductance of transistors $M1$ and $M2$, respectively. From (5), the noise can be analyzed as follows: the thermal noise contribution from R_1 is directly referred to the input, and as the frequency increases, the noise is dominated by terms containing ω^2. Therefore, a low input-referred noise can be achieved by increasing resistor R_1, yielding better TIA noise performance.

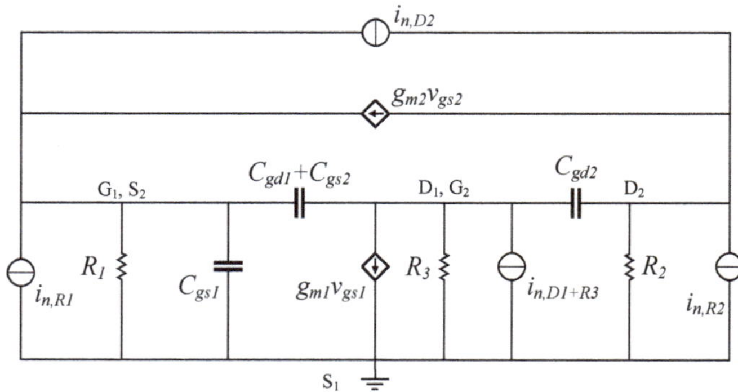

Figure 5. Simplified equivalent circuit of the RGC input stage used for noise analysis.

4. Experimental Results

The resonator and the TIA were both characterized, and were then combined to implement the MEMS-based oscillator. Two test configuration setups shown in Figure 6 were used to characterize the MEMS-based oscillator: (i) the open-loop configuration and (ii) the closed-loop configuration.

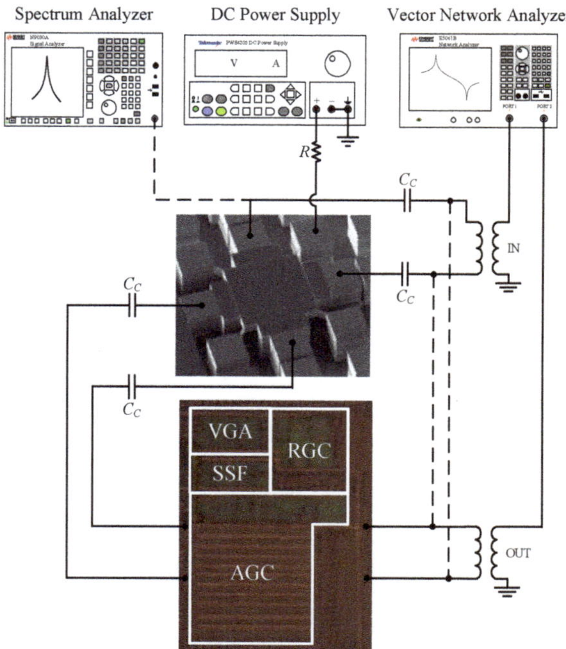

Figure 6. Test setup of the MEMS-based oscillator in open-loop (solid lines) and closed-loop (dashed lines) with micrographs of the TIA and resonator.

4.1. Resonator Characterization

The frequency response of the resonator was measured in differential configuration with the VNA under a vacuum level of 100 mTorr for DC polarization voltages, V_p, of 100 V and 200 V, and for various input power levels starting from -30 dBm up to 0 dBm. Figure 7 shows the transmission characteristic curves normalized to the center frequency of 17.93 MHz. The resonator exhibits a Q-factor of $\sim$890,000, and a peak transmissions of -57 dB and -45 dB for V_p = 100 V and V_p = 200 V, which correspond to motional resistances of 35 kΩ and 8.8 kΩ, respectively. The results at high input power levels indicate spring-hardening non-linear behavior, as the Lamé-mode resonator geometry is aligned with the <100> crystalline silicon orientation [34–36]. Therefore, a positive amplitude–frequency (A–f) coefficient (κ) is associated with this resonator [37].

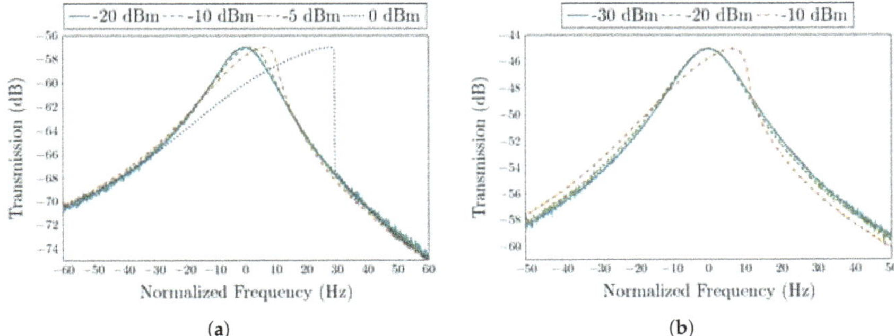

Figure 7. Normalized resonator transmission characteristic curves for various output input amplitude levels for (a) V_p = 100 V and (b) V_p = 200 V.

4.2. Transimpedance Amplifier Characterization

The fully differential TIA is fabricated in a 65 nm CMOS process from TSMC, and consumes only 0.9 mA from a 1-V supply. The total circuit area measures 130 $\times$ 225 μm^2, as shown in Figure 6. To obtain the frequency response of the TIA, S-parameters were measured using a Keysight E5061B VNA in a frequency range from 100 kHz to 1 GHz with an input power level of -45 dBm. Figure 8 shows the transimpedance gain and the 3-dB bandwidth of the TIA, versus two control signals, V_{CTRL_A} and V_{CTRL_BW}. The maximum achievable gain is 98 dBΩ with a bandwidth of 90 MHz. The bandwidth can be extended to 142 MHz when the gain is reduced to 83 dBΩ. Control voltages can be varied independently in such a way that the gain and bandwidth are also independent from each other. As such, as V_{CTRL_BW} varies from 0.35 V to 0.45 V, the maximum gain variation (for the same V_{CTRL_A} value) is $\sim$0.32 dB (as seen in Figure 8a). The motional resistances of 35 kΩ and 8.8 kΩ, extracted from Figure 7 for V_p of 100 V and 200 V, respectively, correspond to 91 dBΩ and 79 dBΩ, respectively, which can be covered by the maximum gain available of the proposed TIA to ensure sufficient gain for oscillation. Figure 9 shows the input-referred current noise of the TIA measured with a Keysight N9030A spectrum analyzer across a 142 MHz bandwidth. At low frequencies, the noise is dominated by the flicker noise, while the input current noise spectrum is flat in the frequency range from $\sim$500 kHz to 142 MHz where the input-referred noise is dominated by the white noise and reaches 15 pA/$\sqrt{\text{Hz}}$. Figure 10 shows the measured transimpedance gain for different input power levels varying from -50 dBm to -35 dBm. The TIA 1-dB compression point was extracted to be of -39 dBm. The performance parameters of the TIA are summarized in Table 1.

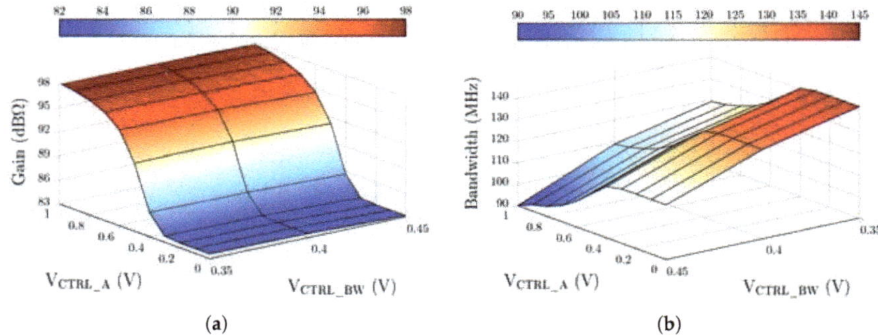

Figure 8. Measured TIA (**a**) gain and (**b**) bandwidth, for different values of V_{CTRL_A} and V_{CTRL_BW}.

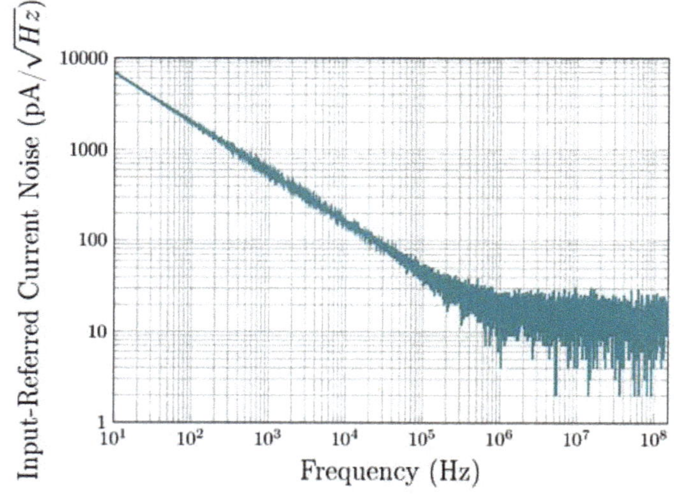

Figure 9. Measured TIA input-referred current noise.

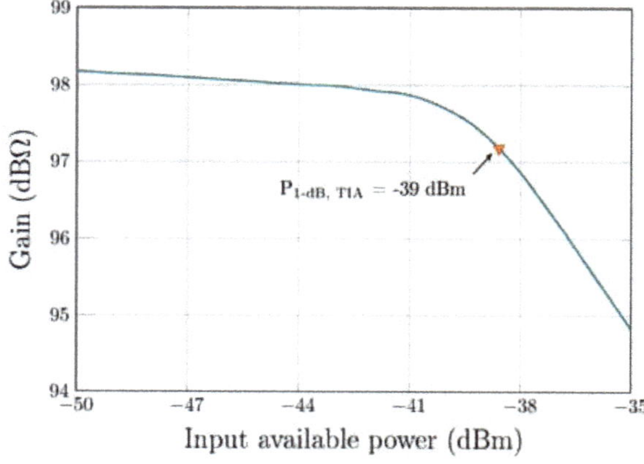

Figure 10. Measured TIA gain for different input power levels, outlining the 1-dB compression point.

Table 1. Performance parameters of the proposed TIA.

Parameter	Meas. Value
Transimpedance gain [dBΩ]	98
Bandwidth [MHz]	142
Input impedance, R_{in} @f_0 [Ω]	89
Output impedance, R_{out} f_0 [Ω]	100
Power supply, V_{DD} [V]	1
Power Consumption, P_{diss} [mW]	0.9
1-dB compression point, $P_{1\text{-dB}}$ [dBm]	−39
Input-referred noise @f_0 $\left[\text{pA}/\sqrt{\text{Hz}}\right]$	14.5
Active area [mm^2]	0.029
Process	65 nm CMOS

4.3. MEMS Oscillator Characterization

4.3.1. Open-Loop Measurements

To confirm that sufficient loop gain was present for the oscillation, the resonator was connected to the TIA in open-loop configuration under vacuum, and the frequency and phase responses were measured using a Keysight E5061B VNA. As illustrated in Figure 6, the input and output ports of the VNA were connected to the differential inputs of the resonator and the differential outputs of the TIA, respectively, through external baluns which are used to convert between single-ended and differential signals. To sustain oscillation in closed-loop, the following conditions are required [37]:

$$\phi_{total} = 0°, \text{ and} \tag{6}$$

$$Z_T \geq R_m + R_{in} + R_{out}, \tag{7}$$

where ϕ_{total} is the total phase shift, R_m is the motional resistance of the resonator, and Z_T, R_{in} and R_{out} are the transimpedance gain, input, and output impedances of the TIA, respectively. In this case, both resonator and TIA must have 0° phase shift. The open-loop gain and phase characteristics are plotted in Figure 11. It is observed that the open-loop gain and phase shift at the resonant frequency of the resonator is higher than 0 dB and equal to 0°, respectively, as formulated in conditions (6) and (7), thus ensuring that oscillation can be sustained in closed-loop. In addition, the loaded Q-factor was measured from the open-loop gain bandwidth to be around 875,000.

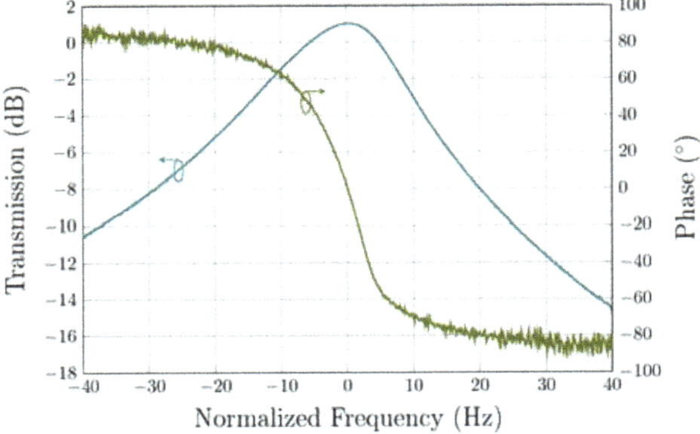

Figure 11. Measured open-loop gain and phase shift of the oscillator loop under vacuum at a polarization voltage of 100 V.

4.3.2. Closed-Loop Measurements

The resonator and TIA were set in a closed-loop configuration (dashed lines in Figure 6) and tested under vacuum to characterize the performance of the oscillator. The expression for oscillator phase noise is given as follows [38]:

$$\mathcal{L}\left(f_m\right) \; = \; \frac{2FkT}{P_0} \times \left[1 + \left(\frac{f_0}{2Q_L\,f_m}\right)^2 \times \left(1 + \frac{f_c}{f_m}\right)\right],$$

(8)

where F represents the noise factor of the amplifier, P_0 is defined as the oscillation power, f_m the offset frequency from the carrier frequency, f_0 represents the carrier frequency, f_c is a constant related to the $1/f$ noise corner of the oscillator and Q_L denotes the loaded Q-factor and is defined as

$$Q_L \; = \; Q_{UL} \times \frac{R_m}{R_m + R_{in} + R_{out}},$$

(9)

where Q_{UL} is the intrinsic Q-factor of the resonator. The phase noise measurements of the oscillator under vacuum are plotted in Figure 12 for polarization voltages of 100 V and 200 V. The near-carrier phase noise at a 10 Hz offset was measured to be approximately of -50 dBc/Hz and of -70 dBc/Hz at polarization voltages of 100 V and 200 V, respectively. At an offset of 1 kHz, the phase noise was measured to be of -120 dBc/Hz at both polarization voltages. At a polarization voltage of 100 V, the TIA flicker noise dominates the close-to-carrier phase noise. However, at a polarization voltage of 200 V, the close-to-carrier phase noise is deteriorated by the resonator non-linearity [24]. Figure 7 clarifies the effect of the polarization voltage on the non-linearity, by comparing the response at 100 V and 200 V polarization. The resonator exhibits significant non-linearity at a -10 dBm signal input and 200 V polarization, whereas, the non-linearity is significantly reduced at 100 V polarization at the same signal level. This results in the phase noise in the close-to-carrier region to be improved by $\sim$20 dB when the polarization voltage is decreased. At farther frequency offsets, the phase noise reaches a floor of -127 dBc/Hz and is dominated by the TIA noise. These phase noise measurements translate in time-domain jitter values. The RMS integrated phase jitter (from 12 kHz to 20 MHz) is of 14 ps. Short-term stability is an important performance criterion of the oscillator and is a measure of its frequency stability. The frequency stability of the resonator is illustrated in Figure 13. The oscillator shows a broadening of the frequency output over a five-minute timespan. Please note that the frequency stability is improved when the AGC is used, from ± 1.3 ppm to ± 0.25 ppm, as this ensures that the non-linearity of the resonator is not exerted.

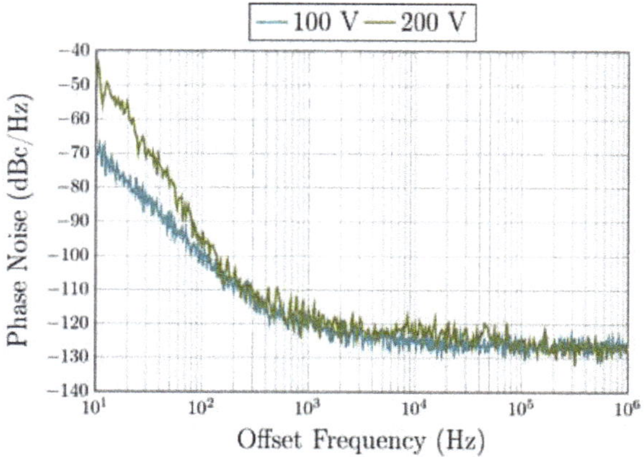

Figure 12. Measured phase noise in vacuum for polarization voltages of 100 V and 200 V.

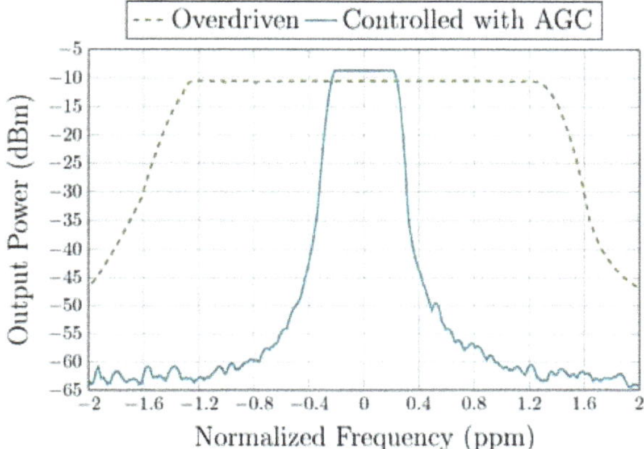

Figure 13. MEMS oscillator signal short time stability at a 17.93 MHz central frequency (averaged over a five-minute timespan) with a polarization voltage of 100 V.

To allow for a representative comparison, two figure-of-merits FOM_1, and FOM_2, are used to evaluate the overall MEMS oscillator performance. Their expressions are respectively given by [19,39]

$$FOM_1 = \mathcal{L}(f_m) - 20\log\left(\frac{f_0}{f_m}\right) + 10\log\left(\frac{P_{\text{diss}}}{1\text{mW}}\right), \tag{10}$$

and

$$FOM_2 = \frac{kT}{\text{PN Floor} \times P_{\text{diss}}} f_0^2 R_m^2, \tag{11}$$

where $\mathcal{L}(f_m)$ is the oscillator phase noise at f_m, a specific offset frequency, f_0 is the center frequency, P_{diss} is the DC power consumption of the oscillator circuit (in mW), PN Floor is the phase noise floor (in dBc/Hz), and R_m is the motional resistance. The calculated FOM_1 and FOM_2 values for different MEMS oscillators based on electrostatic resonators in the literature, both monolithic and non-monolithic, are listed in Table 2. It can be noticed that proposed FOM_2 is used to evaluate the phase noise floor enabled by the TIA while considering the high resonator motional resistance [19]. The calculated FOM_1 and FOM_2 values for different MEMS oscillators based on electrostatic resonators in the literature are listed in Table 2. As can be seen, the MEMS-based oscillator demonstrated in this work has the highest figure-of-merit $|FOM_1|$ when compared to others [19–26], while [25,26] have higher FOM_2, but exhibit significantly lower Q and higher motional resistance, ultimately leading to reduced phase noise performance. These works also present all-CMOS oscillators which impose more limitations on the material choices and resonator fabrication compared to the hybrid approach adopted in this work. Its close-to-carrier phase noise is notably lower because of the low noise of the TIA and the mitigation of the resonator non-linearity. Accordingly, the proposed design enables competitive performance while operating at a low power consumption of 0.9 mW.

Table 2. Performance comparison of the proposed oscillator with the state-of-the-art.

	[19]	[20]	[21]	[22]	[23]	[24]	[25]	[26]	This Work
CMOS technology	0.35 μm	0.18 μm	0.35 μm	0.18 μm	0.35 μm	0.35 μm	0.35 μm	0.35 μm	65 nm
Gap [nm]	1500	200	100	50	80	60	450	900	2000
Center frequency, f_0 [MHz]	20	103	10.92	18	61.2	1.18	3.2	1.23	17.93
Testing condition	vacuum	air	vacuum	vacuum	air	vacuum	vacuum	vacuum	vacuum
Quality factor, Q	160,000	80,000	1092	8000	48,000	3029	2228	1900	889,539
Motional resistance, R_m [kΩ]	65	5	6	76.9	15	700	12,000	16,000	35
Polarization voltage, V_p [V]	26	18	5	2.5	12	45	30	7	100
AGC Circuit	Yes	Yes	No	No	No	No	No	No	Yes
Power supply, V_{DD} [V]	2.5	1.8	3.3	1.8	3.3	2.5	3.3	2.5	1
Power consumption, P_{diss} [mW]	6.9	2.6	0.35	5.9	0.95	1.3	1.21	0.15	0.9
PN @1kHz [dBc/Hz]	−105	−108	−80	−116	−100	−112	−82	−106	−120
PN Floor [dBc/Hz]	−131	−136	−96	−130	−130	−120	−105	−111	−127
FOM$_1$ @1kHz [dB]	−183	−204	−165	−193	−196	−172	−71	−96	−205
FOM$_2$ [Hz^{2}Ω^2]	3.6×10^{17}	1.3×10^{19}	1.9×10^{14}	2.2×10^{18}	1.7×10^{19}	1.2×10^{19}	1.58×10^{20}	1.34×10^{21}	3.8×10^{19}

Monolithic integration typically results in lower parasitic capacitances and consequently lower power dissipation, which is reflected in the FOMs, e.g., FOM$_1$ for [21,24]. On the other hand, it imposes fabrication limitations on the resonator, e.g., thermal budget, materials, and processing steps compatibility with the electronics.

5. Conclusions

This paper presented a MEMS oscillator based on a Lamé-mode capacitive MEMS resonator and a fully differential high-gain TIA. The TIA was fabricated in a TSMC 65 nm CMOS process from TSMC and consumes 0.9 mW. An RGC input stage was used in this work to benefit from high gain, wide bandwidth, and lower input impedance which make it suitable for oscillators based on capacitive MEMS resonators. The TIA can reach a maximum gain of 98 dBΩ and has a bandwidth that is adjustable from 90 to 142 MHz, making it versatile for use with different resonators to attain suitable oscillation conditions. The input-referred current noise of the TIA was measured below 15 pA/$\sqrt{\text{Hz}}$ in the mid-band. The proposed TIA was integrated with an 18-MHz Lamé-mode MEMS resonator to implement a MEMS oscillator. The presented MEMS oscillator achieves a phase noise of −120 dBc/Hz, at a 1-kHz offset and the phase noise floor is of −127 dBc/Hz. The oscillator exhibits a superior figure-of-merit relative to the state-of-the-art, notably in terms of power consumption and phase noise.

Author Contributions: A.B. and M.Y.E. designed the devices, the test setups and performed the experimental testing. F.N. supervised the work and provided expertise. All authors contributed to the writing of the paper.

Funding: This research was funded by the Natural Sciences and Engineering Research Council of Canada (NSERC), the Fonds de recherche du Québec—Nature et technologies (FRQNT), and the Microsystems Strategic Alliance of Quebec (ReSMiQ).

Acknowledgments: The authors would like to thank CMC Microsystems for providing access to design software and enabling chip fabrication.

Conflicts of Interest: The authors declare no conflict of interest.

References

1. Elsayed, M.Y.; Nabki, F. 18-MHz Silicon Lamé Mode Resonators with Corner and Central Anchor Architectures in a Dual-Wafer SOI Technology. *J. Microelectromech. Syst.* **2017**, *26*, 67–74. [CrossRef]
2. Khine, L.; Palaniapan, M.; Wong, W.K. 6Mhz Bulk-Mode Resonator with Q Values Exceeding One Million. In Proceedings of the TRANSDUCERS 2007—2007 International Solid-State Sensors, Actuators and Microsystems Conference, Lyon, France, 10–14 June 2007; pp. 2445–2448.
3. Khine, L.; Palaniapan, M. High-Q bulk-mode SOI square resonators with straight-beam anchors. *J. Micromech. Microeng.* **2009**, *19*, 015017. [CrossRef]

4. Lee, J.Y.; Seshia, A. 5.4-MHz single-crystal silicon wine glass mode disk resonator with quality factor of 2 million. *Sens. Actuators A Phys.* **2009**, *156*, 28–35. [CrossRef]

5. Xu, Y.; Lee, J.E.Y. Mechanically coupled SOI Lamé-mode resonator-arrays: Synchronized oscillations with high quality factors of 1 million. In Proceedings of the 2013 Joint European Frequency and Time Forum International Frequency Control Symposium (EFTF/IFC), Prague, Czech Republic, 21–25 July 2013; pp. 133–136.

6. Zhu, H.; Xu, Y.; Lee, J.E.Y. Piezoresistive Readout Mechanically Coupled Lamé Mode SOI Resonator with Q of a Million. *J. Microelectromech. Syst.* **2015**, *24*, 771–780. [CrossRef]

7. Xereas, G.; Chodavarapu, V.P. Wafer-Level Vacuum-Encapsulated Lamé Mode Resonator with f-Q Product of 2.23×10^{13} Hz. *IEEE Electron Device Lett.* **2015**, *36*, 1079–1081. [CrossRef]

8. Elsayed, M.Y.; Cicek, P.V.; Nabki, F.; El-Gamal, M.N. Bulk Mode Disk Resonator with Transverse Piezoelectric Actuation and Electrostatic Tuning. *J. Microelectromech. Syst.* **2016**, *25*, 252–261. [CrossRef]

9. Thakar, V.; Rais-Zadeh, M. Temperature-compensated piezoelectrically actuated Lamé-mode resonators. In Proceedings of the 2014 IEEE 27th International Conference on Micro Electro Mechanical Systems (MEMS), San Francisco, CA, USA, 26–30 January 2014.

10. Elsayed, M.Y.; Nabki, F. Piezoelectric Bulk Mode Disk Resonator Post-Processed for Enhanced Quality Factor Performance. *J. Microelectromech. Syst.* **2017**, *26*, 75–83. [CrossRef]

11. Lin, Y.W.; Lee, S.; Li, S.S.; Xie, Y.; Ren, Z.; Nguyen, C.C. Series-resonant VHF micromechanical resonator reference oscillators. *IEEE J. Solid-State Circuits* **2004**, *39*, 2477–2491. [CrossRef]

12. Xie, Y.; Li, S.S.; Lin, Y.W.; Ren, Z.; Nguyen, C.C. 1.52-GHz micromechanical extensional wine-glass mode ring resonators. *IEEE Trans. Ultrason. Ferroelectr. Freq. Control* **2008**, *55*, 890–907.

13. Clark, J.; Hsu, W.T.; Abdelmoneum, M.; Nguyen, C.C. High-Q UHF micromechanical radial-contour mode disk resonators. *J. Microelectromech. Syst.* **2005**, *14*, 1298–1310. [CrossRef]

14. Elsayed, M.; Nabki, F.; El-Gamal, M. A 2000°/s dynamic range bulk mode dodecagon gyro for a commercial SOI technology. In Proceedings of the 2011 18th IEEE International Conference on Electronics, Circuits, and Systems, Beirut, Lebanon, 11–14 December 2011.

15. Elsayed, M.Y.; Nabki, F.; El-Gamal, M.N. A combined comb/bulk mode gyroscope structure for enhanced sensitivity. In Proceedings of the 2013 IEEE 26th International Conference on Micro Electro Mechanical Systems (MEMS), Taipei, Taiwan, 20–24 January 2013.

16. Elsayed, M.Y.; Nabki, F.; El-Gamal, M.N. A novel comb architecture for enhancing the sensitivity of bulk mode gyroscopes. *Sensors* **2013**, *13*, 16641–16656. [CrossRef]

17. Elsayed, M.Y.; Nabki, F. Capacitive Lamé mode resonator with gap closing mechanism for motional resistance reduction. In Proceedings of the 2017 Joint Conference of the European Frequency and Time Forum and IEEE International Frequency Control Symposium (EFTF/IFCS), Besancon, France, 9–13 July 2017; pp. 203–205.

18. Elsayed, M.Y.; Nabki, F. 870,000 Q-Factor Capacitive Lamé Mode Resonator with Gap Closing Electrodes Enabling 4.4 kΩ Equivalent Resistance at 50 V. *IEEE Trans. Ultrason. Ferroelectr. Freq. Control* **2019**, *66*, 717–726. [CrossRef] [PubMed]

19. Seth, S.; Wang, S.; Kenny, T.; Murmann, B. A -131-dBc/Hz, 20-MHz MEMS oscillator with a 6.9-mW, 69-kΩ, gain-tunable CMOS TIA. In Proceedings of the 2012 ESSCIRC (ESSCIRC), Bordeaux, France, 17–21 September 2012.

20. Sundaresan, K.; Ho, G.; Pourkamali, S.; Ayazi, F. A Low Phase Noise 100MHz Silicon BAW Reference Oscillator. In Proceedings of the 2006 IEEE Custom Integrated Circuits Conference, San Jose, CA, USA, 10–13 September 2006.

21. Huang, W.L.; Ren, Z.; Lin, Y.W.; Chen, H.Y.; Lahann, J.; Nguyen, C.T.C. Fully monolithic CMOS nickel micromechanical resonator oscillator. In Proceedings of the 2008 IEEE 21st International Conference on Micro Electro Mechanical Systems, Wuhan, China, 13–17 January 2008; pp. 10–13.

22. Chen, T.T.; Huang, J.C.; Peng, Y.C.; Chu, C.H.; Lin, C.H.; Cheng, C.W.; Li, C.S.; Li, S.S. A 17.6-MHz 2.5V ultra-low polarization voltage MEMS oscillator using an innovative high gain-bandwidth fully differential trans-impedance voltage amplifier. In Proceedings of the 2013 IEEE 26th International Conference on Micro Electro Mechanical Systems (MEMS), Taipei, Taiwan, 20–24 January 2013.

23. Lin, Y.W.; Lee, S.; Li, S.S.; Xie, Y.; Ren, Z.; Nguyen, C.T.C. 60-MHz wine-glass micromechanical-disk reference oscillator. In Proceedings of the 2004 IEEE International Solid-State Circuits Conference, San Francisco, CA, USA, 15–19 February 2004; Volume 1, pp. 322–530.

24. Li, M.H.; Chen, C.Y.; Li, C.S.; Chin, C.H.; Li, S.S. A Monolithic CMOS-MEMS Oscillator Based on an Ultra-Low-Power Ovenized Micromechanical Resonator. *J. Microelectromech. Syst.* **2015**, *24*, 360–372. [CrossRef]

25. Uranga, A.; Sobreviela, G.; Riverola, M.; Torres, F.; Barniol, N. Phase-Noise Reduction in a CMOS-MEMS Oscillator Under Nonlinear MEMS Operation. *IEEE Trans. Circuits Syst. I Regul. Pap.* **2017**, *64*, 3047–3055. [CrossRef]

26. Li, M.; Chen, C.; Liu, C.; Li, S. A Sub-150-μWBEOL-Embedded CMOS-MEMS Oscillator with a 138-dBΩUltra-Low-Noise TIA. *IEEE Electron Device Lett.* **2016**, *37*, 648–651. [CrossRef]

27. He, F.; Ribas, R.; Lahuec, C.; Jézéquel, M. Discussion on the general oscillation startup condition and the Barkhausen criterion. *Analog Integr. Circuits Signal Process.* **2009**, *59*, 215–221. [CrossRef]

28. Abdolvand, R.; Bahreyni, B.; Lee, J.E.Y.; Nabki, F. Micromachined Resonators: A Review. *Micromachines* **2016**, *7*, 160. [CrossRef] [PubMed]

29. Pettine, J.; Petrescu, V.; Karabacak, D.M.; Vandecasteele, M.; Crego-Calama, M.; Hoof, C.V. Power-Efficient Oscillator-Based Readout Circuit for Multichannel Resonant Volatile Sensors. *IEEE Trans. Biomed. Circuits Syst.* **2012**, *6*, 542–551. [CrossRef] [PubMed]

30. Sackinger, E.; Guggenbuhl, W. A High-Swing, High-Impedance MOS Cascode Circuit. *IEEE J. Solid-State Circuits* **1990**, *25*, 289–298. [CrossRef]

31. Park, S.; Yoo, H.J. 1.25-Gb/s Regulated Cascode CMOS Transimpedance Amplifier for Gigabit Ethernet Applications. *IEEE J. Solid-State Circuits* **2004**, *39*, 112–121. [CrossRef]

32. Shaeffer, D.; Lee, T. A 1.5-V, 1.5-GHz CMOS low noise amplifier. *IEEE J. Solid-State Circuits* **1997**, *32*, 745–759. [CrossRef]

33. Ogawa, K. Noise Caused by GaAs MESFETs in Optical Receivers. *Bell Syst. Tech. J.* **1981**, *60*, 923–928. [CrossRef]

34. Zhu, H.; Lee, J.E.Y. Orientation dependence of nonlinearity and TCf in high-Q shear-modes of silicon MEMS resonators. In Proceedings of the 2014 IEEE International Frequency Control Symposium (FCS), Taipei, Taiwan, 19–22 May 2014.

35. Zhu, H.; Lee, J.E.Y. Reversed Nonlinear Oscillations in Lamé-Mode Single-Crystal-Silicon Microresonators. *IEEE Electron Device Lett.* **2012**, *33*, 1492–1494. [CrossRef]

36. Zhu, H.; Tu, C.; Lee, J.E.Y. Material nonlinearity limits on a Lamé-mode single crystal bulk resonator. In Proceedings of the 2012 7th IEEE International Conference on Nano/Micro Engineered and Molecular Systems (NEMS), Kyoto, Japan, 5–8 March 2012.

37. Bouchami, A.; Nabki, F. Non-linear modeling of MEMS-based oscillators using an analog hardware description language. In Proceedings of the 2014 IEEE 12th International New Circuits and Systems Conference (NEWCAS), Trois-Rivieres, QC, Canada, 22–25 June 2014.

38. Hajimiri, A.; Lee, T. A General Theory of Phase Noise in Electrical Oscillators. *IEEE J. Solid-State Circuits* **1998**, *33*, 179–194. [CrossRef]

39. Zuo, C.; Spiegel, J.V.D.; Piazza, G. 1.05-GHz CMOS oscillator based on lateral- field-excited piezoelectric AlN contour- mode MEMS resonators. *IEEE Trans. Ultrason. Ferroelectr. Freq. Control* **2010**, *57*, 82–87. [CrossRef] [PubMed]

Article

A Resonant Pressure Microsensor with the Measurement Range of 1 MPa Based on Sensitivities Balanced Dual Resonators

Yulan Lu [1,2], Pengcheng Yan [1,2], Chao Xiang [1,2], Deyong Chen [1,2,*], Junbo Wang [1,2,*], Bo Xie [1,*] and Jian Chen [1]

1   State Key Laboratory of Transducer Technology, Institute of Electronics, Chinese Academy of Sciences, Beijing 100190, China; luyulan15@mails.ucas.ac.cn (Y.L.); yanpengcheng17@mails.ucas.ac.cn (P.Y.); xiangchao115@mails.ucas.ac.cn (C.X.); chenjian@mail.ie.ac.cn (J.C.)
2   University of Chinese Academy of Sciences, Beijing 100049, China
*   Correspondence: dychen@mail.ie.ac.cn (D.C.); jbwang@mail.ie.ac.cn (J.W.); xiebo11@mails.ucas.ac.cn (B.X.); Tel.: +86-010-588-87182 (D.C.); +86-010-588-87191 (J.W.); +86-010-588-87255 (B.X.)

Received: 20 April 2019; Accepted: 13 May 2019; Published: 16 May 2019

Abstract: This paper presents a resonant pressure microsensor with the measurement range of 1 MPa suitable for the soaring demands of industrial gas pressure calibration equipment. The proposed microsensor consists of an SOI layer as a sensing element and a glass cap for vacuum packaging. The sensing elements include a pressure-sensitive diaphragm and two resonators embedded in the diaphragm by anchor structures. The resonators are excited by a convenient Lorentz force and detected by electromagnetic induction, which can maintain high signal outputs. In operation, the pressure under measurement bends the pressure-sensitive diaphragm of the microsensor, producing frequency shifts of the two underlining resonators. The microsensor structures were designed and optimized using finite element analyses and a 4″ SOI wafer was employed in fabrications, which requires only one photolithographic step. Experimental results indicate that the Q-factors of the resonators are higher than 25,000 with a differential temperature sensitivity of 0.22 Hz/°C, pressure sensitivities of 6.6 Hz/kPa, and −6.5 Hz/kPa, which match the simulation results of differential temperature sensitivity of 0.2 Hz/°C and pressure sensitivities of ±6.5 Hz/kPa. In addition, characterizations based on a closed-loop manner indicate that the presented sensor demonstrates low fitting errors within 0.01% FS, high accuracy of 0.01% FS in the pressure range of 20 kPa to 1 MPa and temperature range of −55 to 85 °C, and the long-term stability within 0.01% FS in a 156-day period under the room temperature.

Keywords: resonant pressure microsensor; 1 MPa; dual resonators; sensitivity

1. Introduction

Pressure microsensors are widely used in the fields of industrial metering, aerospace aviation, meteorology, medical, automobiles, and consumer electronics [1–4], which function according to several typical working principles, such as capacitive sensing [5,6], piezoresistive sensing [7,8], piezoelectric sensing [9,10], fiber optical sensing [11,12], and resonant sensing [13,14]. Compared to other types of pressure microsensors, resonant pressure microsensors are featured with high resolutions, high accuracy, long-term stability, and quasi-digital outputs [15,16], which meet the urgent demands of gas pressure calibration equipment in the field of industrial process control.

The resonant pressure microsensor was first introduced by Greenwood in 1984, and is based on electrostatic excitation/capacitive detection [17]. However, the resonator was not packaged in a vacuum, and the microcapacity is hard to detect due to the unwanted interference capacitance, unless in a vacuum environment below 133 Pa. In 1991, Peterson et al. introduced a resonant pressure microsensor based on silicon fusion bonding [18] which achieved vacuum packaged for the resonator. The Q-factor of this sensor was higher than 100,000 and the accuracies were 0.01% FS (Full Scale) in the temperature range of −40 to 125 °C within the pressure measurement range of 10 to 130 kPa. Nevertheless, the vibration directions of the two resonators mentioned above are perpendicular to the pressure-sensitive diaphragm, which would result in energy losses. A laterally driven micromachined resonant pressure sensor was introduced by Welham et al. in 1996 [19]. This sensor uses a hammock resonator as a sensing element and the Q-factor is insensitive to the leakage of cavity gases due to the laterally driven principle. Due to the benefits of being laterally driven, Sun et al. presented a micromachined resonant pressure sensor based on an electrostatic excited/capacitance detected resonator in 2016 [20]. Experiment results showed that the sensor had a pressure sensitivity of ~29 Hz/kPa, a nonlinearity of 0.02% FS, a hysteresis error of 0.05% FS within the pressure measurement range of 20 to 280 kPa, and the temperature sensitivity of the resonator was ~2 Hz/°C in the temperature range of −40 to 80 °C. Meanwhile, Du introduced a resonant pressure microsensor based on an electrostatic excitation/capacitive detection resonator which showed a pressure sensitivity of −8.7 Hz/kPa, a maximum error of 0.0310% FS under the pressure measurement range of 30to 190 kPa and temperature range of 30 to 70 °C [21]. However, those developed sensors included only one resonator which might cause the output frequency strongly sensitive to temperature. Our previous works developed several types of resonant pressure sensors based on dual resonators, such as Luo (2014) [22], Xie (2015) [23], and Shi (2018) [24], whereas the lowest working temperature was constrained, which mainly result from the mismatched sensitivities of two resonators. In addition, the pressure measurement ranges of the developed sensor mentioned above were less than 300 kPa, and much larger pressure measure ranges had not been reported. The compromised pressure measurement range might result from the nonlinearity of frequency responses to the pressure under measurement.

To address these issues, this study presents a resonant pressure microsensor based on dual resonators using convenient magnetic excitation/magnetic detection. This sensor was designed based on balanced temperature and pressure sensitivities within the linear elastic range to explore the low working temperature extremity and enlarge the pressure measurement range. The frequency responses of the resonant sensor were analyzed based on finite element analysis (FEA) simulations. The resonant pressure microsensor was fabricated and experimental characterizations including open-loop tests, closed-loop tests, and long-term tests to validate the design.

2. Design

2.1. Working Principle

The proposed resonant pressure microsensor is shown in Figure 1. The sensor chip with a size of $10 \times 10 \times 0.842 \text{ mm}^3$ consists of a glass cap with eight vias (Φ 600 μm) with a cavity (5 mm × 5 mm × 120 μm) and an SOI (Silicon-on-insulator) wafer with sensing elements. The sensing elements include a pressure-sensitive diaphragm (5 mm × 5 mm × 300 μm) in the handle layer of the SOI wafer, and two H-shaped resonators (two single 1400 μm × 18 μm × 40 μm beams with a connection of 60 μm × 18 μm × 40 μm) in the device layer in the SOI wafer. The doubly clamped H-shaped resonators, which are deployed on the central (resonator I) and border (resonator II) areas of the diaphragm, respectively, are coupled to the diaphragm by anchor structures in the oxide layer of the SOI wafer (see Figure 1a).

In operations, the deflections of the pressure diaphragm build up stresses when the pressure under measurement is applied (see Figure 1b). In this study, the resonator I and resonator II are located in the tensile and compressive area of the diaphragm, which leads to the resonant frequency shifts of resonator I increasing, and those of resonator II decreasing. Note that, the two resonators exhibit the same temperature properties for the same materials and fabrication processes, which means that the differential frequency outputs are immune from temperature variances. For the frequency readout, a permanent magnet was used to provide a uniform magnetic field. An ac signal is fed into one beam of the H-shaped resonator, and the resonator driven by electromagnetic force. On the other hand, the other beam of the resonator can pick up the resonant frequency signals based on the principle of electromagnetic induction (see Figure 1c).

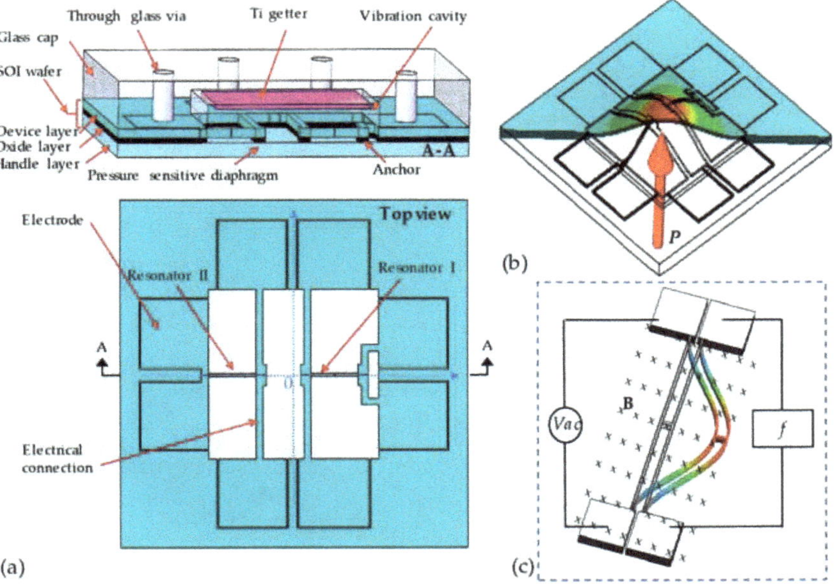

Figure 1. Schematic of the proposed resonant pressure microsensor: (**a**) The proposed pressure microsensor consists of an SOI wafer (including a pressure-sensitive diaphragm, two H-shaped resonators, and eight electrode pads) and a glass cap (including a vibration cavity with getter inside and eight through-glass vias); (**b**) Details of the sensing elements when pressure applied. The pressure-sensitive diaphragm bends under the pressure, which changes the axial stresses of the resonators; (**c**) The resonator is excited based on electromagnetic forces and detected based on electromagnetic inductions.

2.2. FEA Simulations

FEA simulations based on ANSYS software were employed to design and optimize the balanced pressure and temperature sensitivities within the linear elastic range of the resonant pressure microsensor. Multimodels of static structural and modal were used to calculate the intrinsic frequency shifts in response to pressure under measurement variances, and multimodels of steady-state thermal, static structural, and modal were used to calculate the intrinsic frequency shifts in response to surrounding temperature variances. In the simulations, tetrahedral elements were used to mesh the geometrical structures of the microsensor with ~350,000 elements.

The boundary conditions of pressure sensitivity simulations were defined that the edges of the pressure-sensitive diaphragm were constrained to avoid unconstrained movements. In the static structural part, pressures of 100 kPa to 1 MPa were used as the loads. The generated stress distributions within the structures were then used as the loads for the modal part, and the outputs of the simulations were the intrinsic frequency shifts of the resonators in response to the pressure applied.

The initial conditions of temperature sensitivity simulations were set using the reference temperature as 350 °C, which is the bonding temperature of the glass cap and the SOI wafer. In addition, a reference pressure of 100 kPa was introduced in the static structure part for the consideration of convenient comparison to experimental results. In the steady-state thermal simulations, temperatures of −55 to 85 °C were used as the loads and the temperature distributions of the whole structure were transferred to the static structural part. The calculated stresses within the structures were then used as the loads for the modal part, and the outputs of the simulations were the intrinsic frequencies of the resonators in response to temperature variances.

Figure 2a shows the geometry built in the FEA simulations. Figure 2b shows the sensitivity variance in response to the locations of the resonators where the thickness of the pressure-sensitive diaphragm was 300 μm. Figure 2c shows the sensitivity variances in response to the thickness of the pressure-sensitive diaphragm when resonator I was located at +1.2 mm and resonator II was located at −1.8 mm from the center of the pressure-sensitive diaphragm.

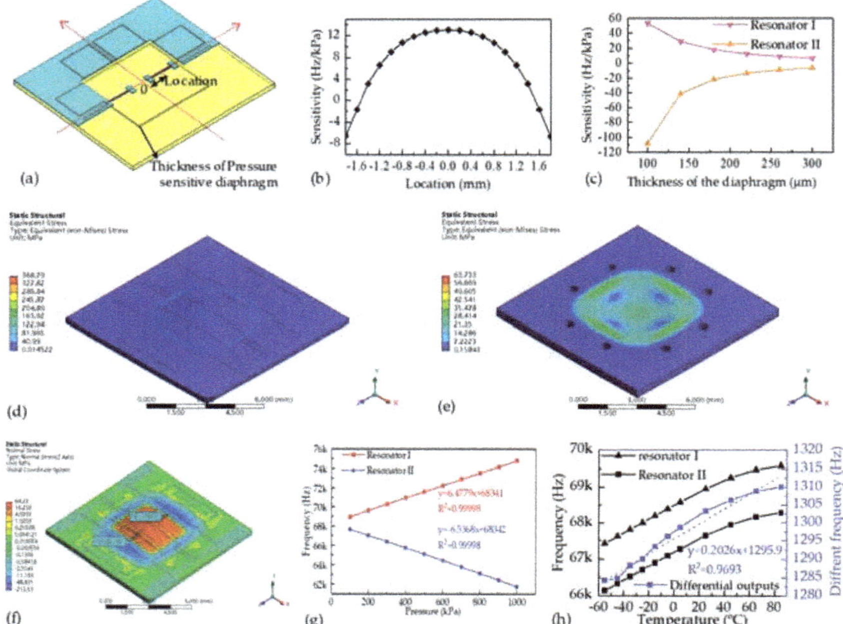

Figure 2. (a) The established geometry using in simulations; (b) The sensitivity variances in response to (b) resonator locations and (c) the thickness of the pressure-sensitive diaphragm. The equivalent stress distributions in (d) the SOI layer and (e) in the glass cap with a pressure of 1 MPa applied. (f) The normal stress distributions along the axial direction of the resonator with a pressure of 1 MPa applied. (g) The intrinsic frequency responses to applied pressure. (h) The intrinsic frequency responses to applied temperature.

From the results of the resonator location simulations and the diaphragm thickness simulations, it can be found that the pressure sensitivities of the microsensor are well-matched when resonator I was positioned at +1.2 mm and resonator II was positioned at −1.8 mm from the center of the pressure-sensitive diaphragm when the thickness of the pressure-sensitive diaphragm was 300 μm. The balanced pressure sensitivities would contribute to temperature compensations due to inversed pressure-sensing properties and similar temperature-sensing properties, which could further extend the pressure measurement range and working temperature range.

Figure 2d–f represent the stress distributions in the developed microsensor under an applied pressure of 1 MPa. The equivalent stresses in the SOI wafer are within the yield strength of 7 GPa (see Figure 2d) and the stresses in the glass cap are within the yield strength of 69 MPa (see Figure 2e), which shows the safety of the microsensor structures. In addition, tensile/compressive stresses are induced in resonator I/II with the values of ±17.1 MPa (see Figure 2f), which indicate that the frequency shifts of the two resonators are the same, but in reversed directions.

Furthermore, the intrinsic frequency shifts as a function of applied pressure and surrounding temperature are shown in Figure 2g,h. The pressure sensitivities of the two resonators were quantified with ±6.5 Hz/kPa and linear coefficients of 0.9999 in the pressure range of 20 kPa to 1 MPa and a reference temperature of 22 °C, and the temperature sensitivity was quantified as 0.2 Hz/°C in the temperature range of −55 to 85 °C with a reference pressure of 100 kPa.

3. Fabrication

Conventional and simplified SOI–MEMS (Micro-Electro-Mechanical System) fabrication processes, which include only one photolithography step, were used to fabricate the proposed resonant pressure microsensor as shown in Figure 3. A 4″ SOI wafer ((100) plane, <100> oriented, p-type, device layer of 40 μm, handle layer of 300 μm, an oxide layer of 2 μm) and a 4″ BF33 glass wafer with a thickness of 500 μm were employed in device fabrications. The main fabrications include deep reactive ion etching, HF releasing, and anodic bonding.

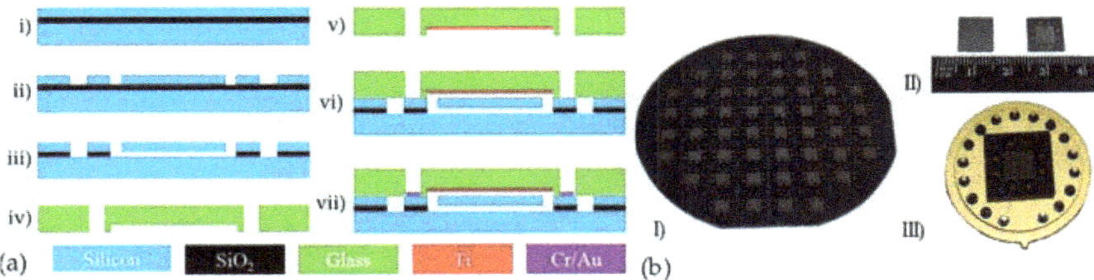

Figure 3. (**a**) The fabrication processes: (i) Cleaning the SOI wafer; (ii) Forming the sensing elements; (iii) Releasing the resonators; (iv) Forming the through-glass vias and cavity in glass wafer; (v) Sputtering Ti as getter; (vi) Anodic bonding; (vii) Cr/Au metallization. (**b**) The fabrication results: (I) Wafer after metallization; (II) Sensor chips after dicing; (III) Packaging prototype.

The SOI wafer was immersed in deionized water and dried with pure N_2 gas after being cleaned by piranha etchant to remove the organic molecules and boiled deionized water to remove soluble ions (see Figure 3a(i)). Then, using patterned photoresist as a mask, the exposed device layer of the SOI wafer was etched through to form the resonators (see Figure 3a(ii)). Note that the handle layer of the SOI wafer remains unetched to form a pressure-sensitive diaphragm with a thickness of 300 μm. After that, the underlying SiO_2 layer of the resonators was removed by immersing the SOI wafer in an HF solution in a time-controlled manner (see Figure 3a(iii)).

The glass wafer was cleaned using the same processes of the SOI wafer. The through-glass vias for electrical connections and the cavities for housing the vibration of the resonators in the BF33 glass wafer were drilled by laser in a program-controlled manner (see Figure 3a(iv)). Then, Ti was sputtered on the cavity as the getter material for gas absorption during next anodic bonding process (see Figure 3a(v)).

After finishing the fabrications of the SOI and the glass wafers, anodic bonding was utilized to form vacuum encapsulations for resonators where the voltage, and tool pressure and temperature of anodic bonding were set at 600 V, 1000 mbar, and 350 °C, respectively (see Figure 3a(vi)). Then, Cr/Au (200 Å/1000 Å) films were sputtered on the through-glass vias to form electrical connections by using a hard mask (see Figure 3a(vii)). The fabrication results were shown in Figure 3b, including the microsensor wafer after Cr/Au metallization (see Figure 3b(I)), the front/back views of the microsensor chips after dicing (see Figure 3b(II)), and the microsensor after packaging (see Figure 3b(III)).

4. Experimental Characterizations

The open-loop circuit was used to quantify the intrinsic frequencies, phase shifts, and Q-factors of the two resonators of the proposed microsensor under an atmospheric pressure of ~100 kPa and room temperature of ~25 °C as shown in Figure 4a. A network analyzer was used to supply an ac voltage to one beam of the resonator and pick up the voltage induced by the other beam of the resonator. The intrinsic frequency of the resonator I was quantified as 68.951 kHz with the phase shift of ~180° and Q-factor of 27174 (see Figure 4b). Besides, the intrinsic frequency of the resonator II was quantified as 67.650 kHz with the phase shift of ~180° and Q-factor of 25612 (see Figure 4c). The mismatches of the Q-factors mainly result from the different structural damping effects of the two resonator structures.

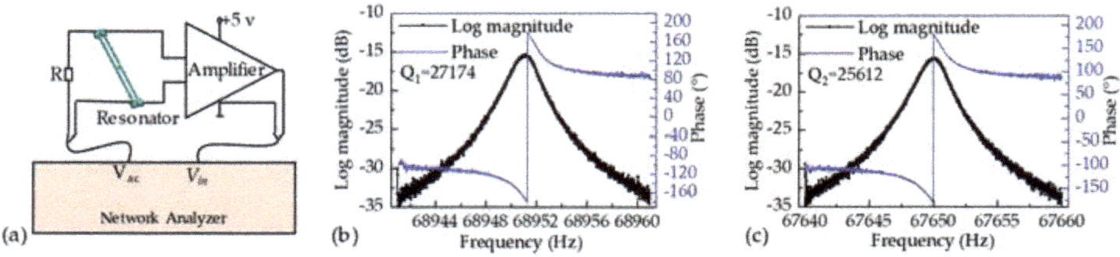

Figure 4. (a) Schematic of the open-loop measurements for the resonators of the proposed microsensor. The intrinsic frequencies, phase shifts, and Q-factors of resonator I (b) and resonator II (c) under an atmospheric pressure of ~100 kPa and room temperature of ~25 °C.

In order to further characterize the performances of the proposed microsensor, a closed-loop circuit producing the self-oscillation signal was developed, as shown in Figure 5a. An amplifier was used to amplify the voltage induced by the vibrations of the resonator, and the amplified voltage was then depressed and sent to the driving beam of the resonator for exciting the resonator in a closed-loop manner. In order to maintain the stable vibrations of the resonator, an automatic gain- controlled module, including a bandpass filter, a comparator, and a field effect transistor, was introduced in the closed-loop circuit.

A pressure calibrator (PPC4, FLUCK) and a temperature chamber (SU-262, ESPEC) were employed to provide the pressure under measurement and surrounding temperature during the characterization processes. In this study, the microsensor was characterized within a pressure range from 20 kPa to 1 MPa and a temperature range from −55 to 85 °C.

Figure 5b shows the intrinsic frequencies of the two resonators as a function of pressure under measurement at 25 °C, producing the pressure sensitivities and linearly dependent coefficient of +6.6 Hz/kPa, 0.99998 for resonator I and −6.5 Hz/kPa, 0.99998 for the resonator II. Figure 5c shows the intrinsic frequencies of the two resonators as a function of surrounding temperature with a pressure of 100 kPa applied, producing the differential temperature sensitivity of 0.22 Hz/°C in the temperature range of −55 to 85 °C. The two results validated the accuracies of the FEA simulations.

Considering the wide working temperature range of the developed microsensor, temperature compensation was performed in this study. As mentioned above, the frequency of the resonators can be expressed as functions of pressure and temperature, which means that the pressure P can be obtained by the frequencies of resonators (f_1 and f_2) based on mathematical translation [25]. Then, the pressure P can be expressed as

$$P = a_0 + a_1 f_1 + a_2 f_2 + a_3 f_1^2 + a_4 f_1 f_2 + a_5 f_2^2 + a_6 f_1^3 + a_7 f_1^2 f_2 + a_8 f_1 f_2^2 + a_9 f_2^3, \tag{1}$$

where $a_0, \ldots, a_9$ are temperature compensation coefficients.

Figure 5d shows the fitting errors of the proposed microsensor in the full pressure and temperature ranges following the custom temperature compensation algorithm, producing the compensation errors within ±76 Pa with corresponding ±0.01% FS, which indicates that the developed microsensor is stable enough under the heat conditions from −55 to 85 °C. Figure 5e shows the measurement errors of the microsensor under the surrounding temperatures of −55, 25, and 85 °C, which demonstrates a high accuracy with quantified measurement errors within ±67 Pa with the corresponding ±0.01% FS. These results validate the functionalities of the balanced pressure sensitivities in extending the pressure and temperature ranges.

Furthermore, the long-term measurements for the actual atmospheric pressure of the developed microsensor in room temperature were conducted for a 156-day period. As shown in Figure 5f, the pressure deviations between the developed microsensor and the PACE5000 Pressure Calibrator (GE Druck, USA) are less than ±20 Pa with the corresponding ±0.002% FS. These results validated the stabilities of the resonant pressure microsensor.

Table 1 shows the comparison of the proposed microsensor to previously reported counterparts, which indicated the proposed microsensor demonstrate higher pressure measurement range, lower working temperature extremity, lower temperature sensitivity, equivalent Q-factor, and equivalent accuracy.

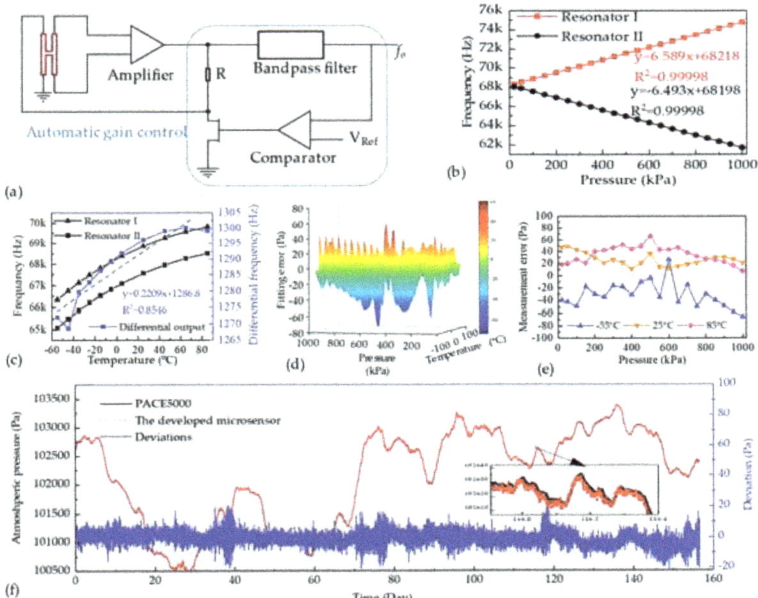

Figure 5. (**a**) Schematic of the closed-loop circuit. (**b**) The frequency responses as functions of the applied pressure in room temperature. (**c**) The frequency responses as functions of the surrounding temperature with the reference pressure of 100 kPa. (**d**) The fitting errors of the developed microsensor within the full pressure range of 20 kPa to 1 MPa and temperature range of −55 to 85 °C. (**e**) The measurement errors to the standard pressure of 20 kPa to 1 MPa under the surrounding temperature of −55, 25, and 85 °C. (**f**) The comparisons of the developed microsensor and the PACE5000 pressure calibrator in actual atmospheric pressure measurement.

Table 1. Comparisons of microsensor performances.

Reference	Q-Factor	Pressure Range	Temperature Range	Temperature Sensitivity	Accuracy
Peterson [18]	>100,000	10~130 kPa	−40~125 °C	~2 Hz/°C	0.01% FS
Welham [19]	50 in air	0~350 kPa	-	-	-
Sun [20]	10,000	20~280 kPa	−40~80 °C	2 Hz/°C	0.05% FS
Du [21]	20,000	30~190 kPa	30~75 °C	-	0.03% FS
Luo [22]	22,000	50~100 kPa	−40~70 °C	−0.33 Hz/°C	0.02% FS
Xie [23]	11,000	50~110 kPa	−40~70 °C	-	0.02% FS
Shi [24]	10,000	10~150 kPa	−35~85 °C	−0.30 Hz/°C	0.01% FS
This sensor	>25,000	20~1000 kPa	−55~85 °C	+0.22 Hz/°C	0.01% FS

5. Conclusions

A resonant pressure microsensor with the measurement range of 1 MPa was presented, where FEA simulations were used to design the sensor structure. The microsensor was fabricated based on conventional and simplified fabrication processes, which need only one photolithography step. The experimental results show that the sensitivities of the two resonators of the developed sensor are 6.6 Hz/kPa and −6.5 Hz/kPa, both with linearly dependent coefficients of 0.99998, which match the simulation results. The differential temperature sensitivity of the developed sensor is 0.22 Hz/°C, which is a benefit resulting from the design of the dual resonator. Moreover, the Q-factors of the two resonators were quantified at higher than 25,000. Further characterizations based on a closed-loop manner indicate that the developed sensor demonstrates low fitting errors within 0.01% FS, and low measurement errors within 0.01% FS under the pressure range of 20 kPa to 1 MPa in a temperature range of −55 to 85 °C. The long-term stability of the proposed microsensor was validated by comparing the measurements of the actual atmospheric pressure with PACE5000 calibrator in a 156-day period, exhibiting a lower deviation within 0.002% FS.

Author Contributions: Defined the project, J.W., D.C, B.X. and Y.L.; designed the device, Y.L. and B.X.; fabricated the devices, Y.L. and P.Y.; characterized the devices, Y.L. and C.X.; drafted the manuscript Y.L. and J.C.

Funding: The authors acknowledge financial supports from the National Key Research and Development Program (2018YFF01010400), the National Natural Science Foundation of China (Grant No. 61825107 and 61431019).

Conflicts of Interest: The authors declare no conflict of interest.

References

1. Eaton, W.P.; Smith, J.H. Micromachined pressure sensors: Review and recent developments. *Smart Mater. Struct.* **1997**, *6*, 530–539. [CrossRef]
2. Beeby, S.P.; Ensel, G.; Kraft, M.; White, N.M. *MEMS Mechanical Sensors*; Artech House Inc.: London, UK, 2008; pp. 97–112.
3. Wang, J.B.; Chen, D.Y.; Xie, B.; Chen, J.; Zhu, L.; Lu, Y.L. A micromachined silicon resonant pressure sensor. In *Micro Electro Mechanical Systems*; Huang, Q.A., Ed.; Springer: Singapore, 2017; pp. 1–34.
4. Greenwood, J.; Wray, T. High accuracy pressure measurement with a silicon resonant sensor. *Sens. Actuators A Phys.* **1993**, *37*, 82–85. [CrossRef]
5. Wygant, I.O.; Kupnik, M.; Khuri-Yakub, B.T. An analytical model for capacitive pressure transducers with circular geometry. *J. Microelectromech. Syst.* **2018**, *27*, 448–456. [CrossRef]
6. Unigarro, E.; Bohorquez, J.C.; Achury, A.; Ramirez, F.; Sacristan, J.; Segura-Quijano, F. Differential capacitive pressure sensor design based on standard CMOS. *Electron. Lett.* **2017**, *53*, 741–742. [CrossRef]
7. Mosser, V.; Suski, J.; Goss, J.; Obermeier, E. Piezoresistive pressure sensors based on polycrystalline silicon. *Sens. Actuators A Phys.* **1991**, *28*, 113–132. [CrossRef]
8. Li, C.; Xie, J.B.; Cordovilla, F.; Zhou, J.Q.; Jagdheesh, R.; Ocana, J.L. Design, fabrication and characterization of an annularly grooved membrane combined with rood beam piezoresistive pressure sensor for low pressure measurements. *Sens. Actuators A Phys.* **2018**, *17*, 525–536. [CrossRef]
9. Yang, F.; Kong, D.R.; Kong, L. Accurate measurement of high-frequency blast waves through dynamic compensation of miniature piezoelectric pressure sensors. *Sens. Actuators A Phys.* **2018**, *280*, 14–23. [CrossRef]
10. Xu, F.J.; Ma, T.H. Modeling and Studying Acceleration-Induced Effects of Piezoelectric Pressure Sensors Using System Identification Theory. *Sensors* **2019**, *19*, 1052. [CrossRef]
11. Shi, H.; Gong, J.Z.; Cook, P.R.; Arkwright, J.W.; Png, G.M.; Lambert, M.F.; Zecchin, A.C.; Simpson, A.R. Wave separation and pipeline condition assessment using in-pipe fibre optic pressure sensors. *J. Hydroinform.* **2019**, *21*, 371–379. [CrossRef]
12. Li, W.W.; Liang, T.; Jia, P.G.; Lei, C.; Hong, Y.P.; Li, Y.W.; Yao, Z.; Liu, W.Y.; Xiong, J.J. Fiber-optic Fabry-Perot pressure sensor based on sapphire direct bonding for high-temperature applications. *Appl. Opt.* **2019**, *58*, 1662–1666. [CrossRef]

13. Ikeda, K.; Kuwayama, H.; Kobayashi, T.; Watanabe, T.; Nishikawa, T.; Yoshida, T.; Harada, K. 3-dimensional micromachining of silicon pressure sensor integrating resonant strain-gauge on diaphragm. *Sens. Actuators A Phys.* **1990**, *23*, 1007–1010. [CrossRef]

14. Harada, K.; Ikeda, K.; Kuwayama, H.; Murayama, H. Various applications of resonant pressure sensor chip based on 3-D micromachining. *Sens. Actuators A Phys.* **1999**, *73*, 261–266. [CrossRef]

15. Christopher, J.W.; Greenwood, J.; Bertioli, M.M. A high accuracy resonant pressure sensor by fusion bonding and trench etching. *Sens. Actuators A Phys.* **1999**, *76*, 298–304.

16. Kinnell, P.K.; Craddock, R. Advances in Silicon Resonant Pressure Transducers. *Procedia Chem.* **2009**, *1*, 104–107. [CrossRef]

17. Greenwood, J.C. Etched silicon vibrating sensor. *J. Phys. E Sci. Instrum.* **1984**, *17*, 650–652. [CrossRef]

18. Petersen, K.; Pourahmadi, F.; Brown, J.; Parsons, P.; Skinner, M.; Tudor, M. Resonant Beam Pressure Sensor Fabricated with Silicon Fusion Bonding. In Proceedings of the Transducers '91, International Conference on Solid State Sensors and Actuators, San Francisco, CA, USA, 24–27 June 1991; pp. 664–667.

19. Welham, C.J.; Gardner, J.W.; Greenwood, J. A laterally driven micromachined resonant pressure sensor. *Sens. Actuators A Phys.* **1996**, *52*, 86–91. [CrossRef]

20. Sun, X.D.; Yuan, W.Z.; Qiao, D.Y.; Sun, M.; Ren, S. Design and Analysis of a New Tuning Fork Structure for Resonant Pressure Sensor. *Micromachines* **2016**, *7*, 148. [CrossRef]

21. Du, X.H.; Liu, Y.F.; Li, A.L.; Zhou, Z.; Sun, D.H.; Wang, L.Y. Laterally Driven Resonant pressure sensor with etched silicon dual diaphragms and combined beams. *Sensors* **2016**, *16*, 158. [CrossRef]

22. Luo, Z.Y.; Chen, D.Y.; Wang, J.B.; Li, Y.N.; Chen, J. A High-Q Resonant Pressure Microsensor with Through-Glass Electrical Interconnections Based on Wafer-Level MEMS Vacuum Packaging. *Sensors* **2014**, *14*, 24244–24257. [CrossRef]

23. Xie, B.; Xing, Y.H.; Wang, Y.S.; Chen, J.; Chen, D.Y.; Wang, J.B. A Lateral Differential Resonant Pressure Microsensor Based on SOI-Glass Wafer-Level Vacuum Packaging. *Sensors* **2015**, *15*, 24257–24268. [CrossRef]

24. Shi, X.Q.; Lu, Y.L.; Xie, B.; Li, Y.D.; Wang, J.B.; Chen, D.Y.; Chen, J. A resonant pressure microsensor based on double-ended tuning fork and electrostatic excitation/piezoresistive detection. *Sensors* **2018**, *18*, 2494. [CrossRef]

25. Li, Y.N.; Wang, J.B.; Luo, Z.Y.; Chen, D.Y.; Chen, J. A Resonant Pressure Microsensor Capable of Self-Temperature Compensation. *Sensors* **2015**, *15*, 10048–10058. [CrossRef]

 *sensors*

Article

Complementary Metamaterial Sensor for Nondestructive Evaluation of Dielectric Substrates

Tanveer ul Haq [1], **Cunjun Ruan** [1,2,*], **Xingyun Zhang** [1] and **Shahid Ullah** [1]

[1] School of Electronic and Information Engineering, Beihang University, Beijing 100191, China; tanveerulhaq@buaa.edu.cn (T.u.H.); luckyzhang@buaa.edu.cn (X.Z.); shahidkhan@buaa.edu.cn (S.U.)

[2] Beijing Key Laboratory for Microwave Sensing and Security Applications, Beihang University, Beijing 100191, China

* Correspondence: ruancunjun@buaa.edu.cn; Tel.: +86-135-0120-5336

Received: 22 April 2019; Accepted: 30 April 2019; Published: 7 May 2019

Abstract: In this paper, complementary metamaterial sensor is designed for nondestructive evaluation of dielectric substrates. The design concept is based on electromagnetic stored energy in the complementary circular spiral resonator (CCSR), which is concentrated in small volume near the host substrate at resonance. This energy can be employed to detect various electromagnetic properties of materials under test (MUT). Effective electric permittivity and magnetic permeability of the proposed sensor is extracted from scattering parameters. Sensitivity analysis is performed by varying the permittivity of MUT. After sensitivity analysis, a sensor is fabricated using standard PCB fabrication technique, and resonance frequency of the sensor due to interaction with different MUT is measured using vector network analyzer (AV3672series). The transcendental equation is derived for the fabricated sensor to calculate relative permittivity for unknown MUTs. This method is very simple and requires calculating only the resonant frequency, which reduces the cost and computation time.

Keywords: complementary metamaterial sensor; CCSR; nondestructive evaluation; material under test; permittivity; transcendental equation

1. Introduction

Recently there has been increased growth in utilization of microwave sensors to improve quality assurance in various fields like food [1], healthcare [2], agriculture [3], and environment [4]. Advantages like low cost, ease of fabrication, robust design, and integration with other microwave devices are the main reasons for the popularity of microwave sensors based on complementary metamaterials (MTMs) [5]. Complementary MTMs are manufactured artificially and their properties depend on structure and orientation rather than composition. The first complementary MTM structure was introduced by F. Falcone et al. in 2004 [6] by applying duality argument on split ring resonator (SRR), which was introduced by Pendry et al. as a resonant magnetic particle in 1999 [7]. SRR consisted of two concentric copper rings having a small gap between them and split in each ring to support resonance. SRR acts as an *LC* resonator where the circumference of metallic rings provide inductance while the gap between the rings and splits provides the capacitance. The resonance frequency of SRR depends on the size of splits, diameter, and gap between the rings. Meanwhile, complementary split ring resonator (CSRR) is a negative image of SRR, which can be obtained by etching out SRR from a metallic plate. The intrinsic circuit and excitation model for isolated and coupled SRR and CSRR structures have already been discussed in Ref. [8].

Microwave sensors based on SRRs are usually magnetically coupled with the microstrip transmission line, therefore, these structures are etched on the top layer of microwave sensor near the microstrip line [9]. At the resonance frequency of SRRs, an electric field appears near the narrow split region, which can be employed for liquid characterization [10] and biomedical sensing [11]. SRRs-based sensors are not suitable for microwave sensing of large samples, because of the narrow fringing electric fields. This problem has been solved by utilizing CSRRs in place of SRRs for the measurement of dielectric samples with large dimensions [12]. CSRRs are usually electrically coupled with the microstrip transmission line and etched in the ground plane [13]. At resonance of CSRRs, large fringing electric fields appear in the ground plane, which can be employed to evaluate dielectric thickness [14] and loss tangent [15]. CSRRs have also been used for the synthesis of highly sensitive sensors for concentration of fluid materials [16], and identification of different liquids from a mixture [17].

In this paper, we are using a complementary circular spiral resonator (CCSR) to design a microwave sensor for nondestructive evaluation of dielectric substrates. Electric field concentration, effective electric permittivity, and magnetic permeability of the proposed sensor are calculated numerically. A sensor is fabricated and the magnitude of transmission coefficient (S_{21}) is measured using vector network analyzer (AV3672 series). Five dielectric materials (MUTs) are placed on the fabricated sensor and S_{21} of the sensor is measured due to interaction with these MUTs. This measured data is used to derive a transcendental equation for the sensor. The sensor design is explained in Section 2. Sensitivity analysis is performed in Section 3. Fabrication and measurement results are discussed in Section 4. The transcendental equation is formulated in Section 5 and the research is concluded in Section 6.

2. Sensor Design

The proposed sensor is based on 1 mm thick FR4 substrate, which has two copper layers of 35 μm each. The upper copper layer consists of a microstrip transmission line while the bottom copper layer has a complementary circular spiral resonator (CCSR) as shown in Figure 1. The dielectric constant (ε_{re}) and characteristic impedance (Z_c) of microstrip transmission line are calculated using the following equations [18]:

$$\varepsilon_{re} = \frac{\varepsilon_r + 1}{2} + \frac{\varepsilon_r - 1}{2}\left(1 + 12\frac{h}{w}\right)^{-0.5} \tag{1}$$

$$Z_c = \frac{\eta}{\sqrt{\varepsilon_{re}}}\left\{\frac{w}{h} + 1.393 + 0.677\ln\left(\frac{w}{h} + 1.4444\right)\right\}^{-1} \tag{2}$$

where $\varepsilon_r = 4.4$ is permittivity of dielectric substrate and $\eta = 120\pi\ \Omega$ is impedance of wave in free space, $w = 3$ mm is width of the microstrip transmission line, $h = 1$ mm is height of FR4 epoxy substrate. The proposed sensor is simulated in ANSYS Electronics Desktop 2018, which includes a direct link to a high frequency structure simulator (HFSS). The proposed sensor is excited with electrical polarization in z direction, magnetic polarization in x direction, wave vector in y direction, and the simulation conditions are given in Table 1.

Table 1. ANSYS High Frequency Structure Simulator (HFSS) Simulation Condition.

Analysis Area	Size	$25 \times 30 \times 50$ mm^3
	Boundary Condition	Radiation
Cells	Number	14,201
	Shape	Tetrahedron
Feed		Wave port (50 Ω)
Solution Type		Driven Model
Convergence condition determination		Maximum number of passes; 20 Maximum delta S; 0.02

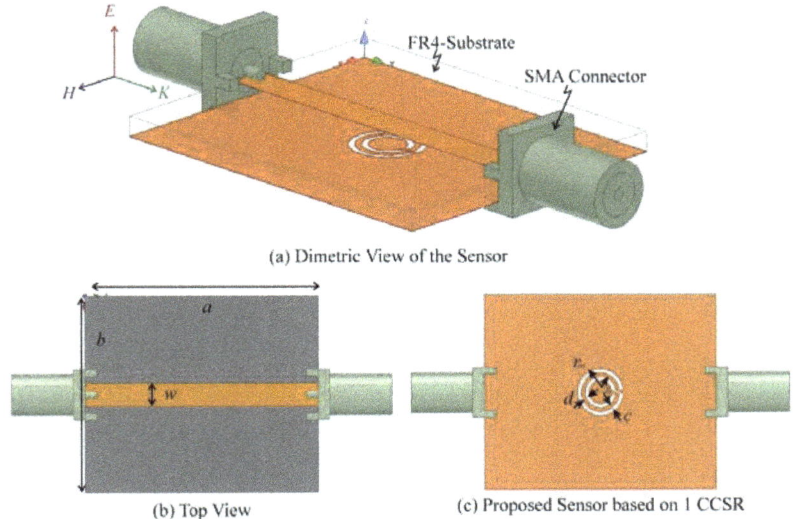

Figure 1. (**a**) Sensor design based on FR4 epoxy substrate (h = 1 mm). (**b**) Top view of the sensor (a = 30 mm, b = 25 mm, w = 3 mm). (**c**) Bottom view of the sensor (r_{ext} = 3 mm, r_0 = 1.5 mm, $c = d = 0.5$ mm).

The magnitude and phase of simulated transmission (S_{21}) and reflection (S_{11}) coefficients are shown in Figure 2; Figure 3, respectively. The fundamental resonance frequency of the sensor is 2.32 GHz with notch depth −16.54 dB. Scattering parameters are used to extract effective electric permittivity (ε) and magnetic permeability (μ) of the proposed sensor using the following relations [19]:

$$\varepsilon = \frac{n}{z} \tag{3}$$

$$\mu = nz \tag{4}$$

where n is refractive index which can be calculated using following equation:

$$n = \frac{1}{kt}\left\{ \mathrm{Im}\left[\ln\left(e^{inkh}\right)\right] - i\mathrm{Re}\left[\ln\left(e^{inkh}\right)\right] \right\} \tag{5}$$

where,

$$e^{inkh} = \frac{S_{21}}{1 - S_{11}R_{01}} \tag{6}$$

where,

$$R_{01} = \frac{z-1}{z+1} \tag{7}$$

where z is the complex impedance which can be calculated using following equation:

$$z = \pm\sqrt{\frac{(1+S_{11})^2 - S_{21}^2}{(1-S_{11})^2 - S_{21}^2}} \tag{8}$$

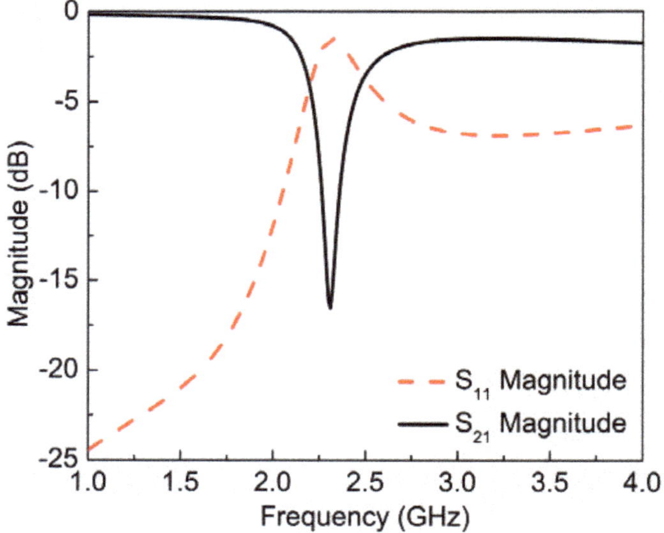

Figure 2. Magnitude of simulated reflection (S_{11}) and transmission (S_{21}) coefficient for the proposed sensor. Resonance frequency of sensor is 2.32 GHz with notch depth −16.54 dB.

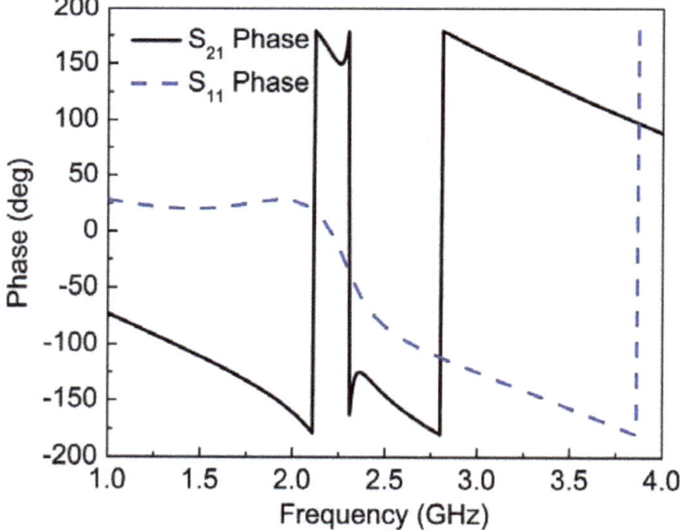

Figure 3. Phase of simulated reflection (S_{11}) and transmission (S_{21}) coefficient for the proposed sensor. A sudden change in phase occurs at resonance frequency.

The extracted plot of electric permittivity for the proposed sensor is plotted in Figure 4 and it shows negative permittivity from 2.10 GHz to 2.12 GHz, and the maximum value of negative permittivity is −3.57 at 2.11 GHz. The extracted plot of magnetic permeability for the proposed sensor is plotted in Figure 5 and it shows negative permeability from 2.27 GHz to 2.42 GHz, and the maximum value of negative permeability is −199.2 at 2.30 GHz. From Figures 4 and 5, it is clear that the proposed sensor is giving negative values of permittivity and permeability, which is a property of metamaterials. The values of electric permittivity and magnetic permeability are dimensionless as they are relative to free space and can vary with the orientation of the sensor, temperature of environment, and molecular structure of the material. The number of CCSR structures in the ground plane of the proposed sensor is increased and their effect on resonance frequency, notch depth, and bandwidth are shown in Figure 6 and tabulated in Table 2. By increasing the number of CCSR in the ground plane, there is no effect on resonance frequency, but notch depth and bandwidth increase and the transmission curve becomes sharper. Since the resonance frequency of each CCSR structure is the same, there is no effect on resonance frequency, but each structure contributes to the notch depth. Due to the increase in notch depth, the bandwidth also increases. Figure 7 shows the electric field distribution at resonance frequency of the proposed sensor based on one, two, three, and four CCSR structures in the ground plane. The maximum magnitude of electric field is 2.31×10^5 V/m for one CCSR in the ground plane and it is concentrated inside the inner ring. For two, three, and four CCSRs, the same electric field distributes among two structures and the maximum magnitude of electric field is 2.20×10^5 V/m, 2.09×10^5 V/m, and 2.0×10^5 V/m, respectively.

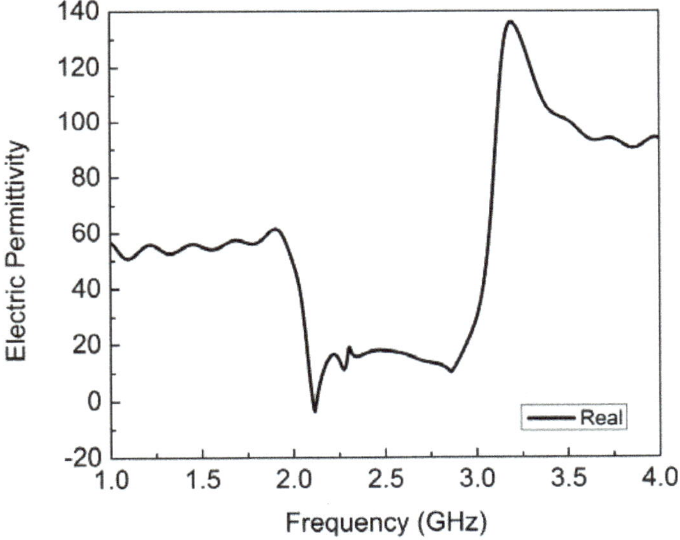

Figure 4. Extracted plot of electric permittivity for proposed sensor, showing negative permittivity from 2.10 GHz to 2.12 GHz, and maximum value of negative permittivity is −3.57 at 2.11 GHz.

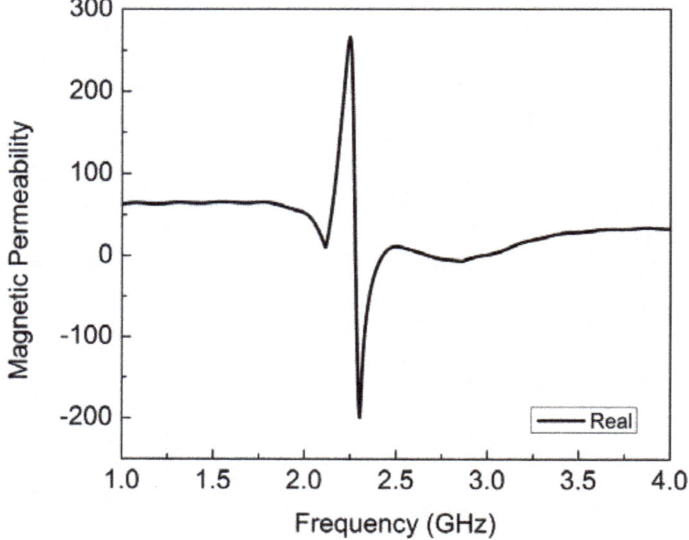

Figure 5. Extracted plot of magnetic permeability for proposed sensor, showing negative permeability from 2.27 GHz to 2.42 GHz, and maximum value of negative permeability is −199.2 at 2.30 GHz.

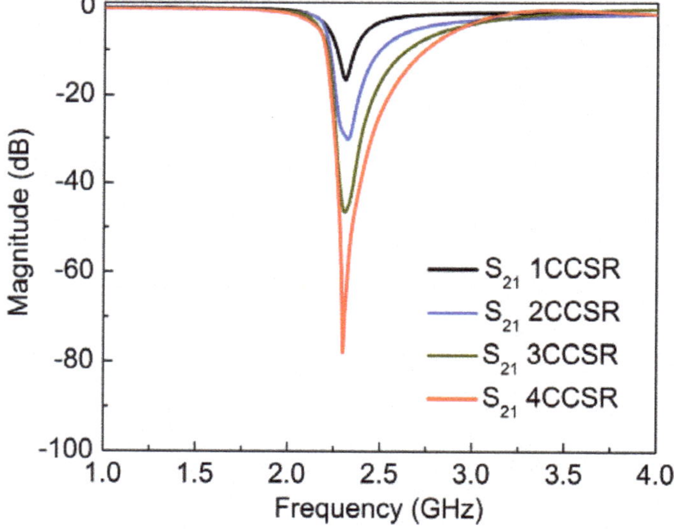

Figure 6. Simulated transmission coefficient of proposed sensor with one, two, three, and four complementary circular spiral resonator (CCSR) in the ground plane separated 1 mm apart. By increasing number of CCSR in the ground plane, notch depth and bandwidth increase but resonance frequency remains the same.

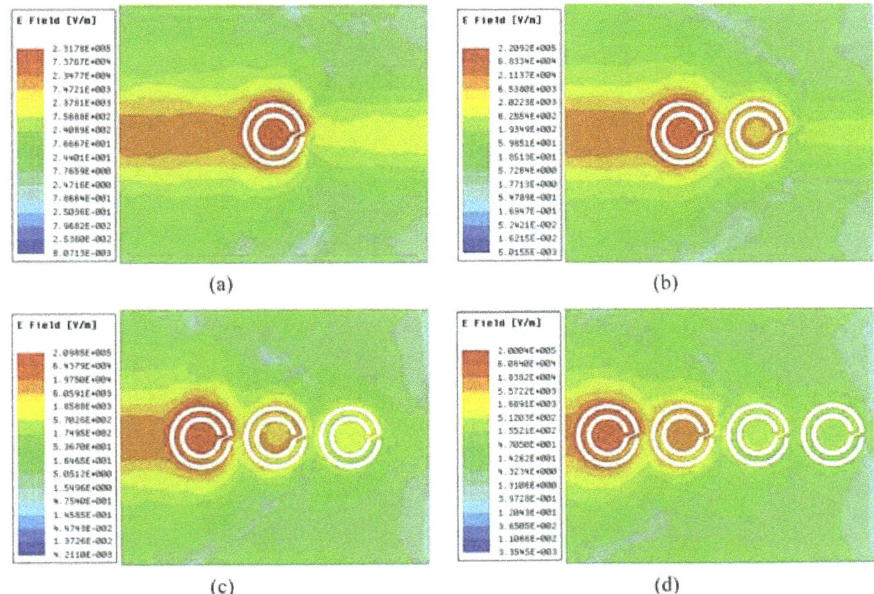

(a)   (b)

(c)   (d)

Figure 7. Distribution of electric field at resonance of proposed sensor based on (**a**) one CCSR, (**b**) two CCSR, (**c**) three CCSR, and (**d**) four CCSR.

Table 2. Effect of CCSR number on resonance frequency, notch depth, and bandwidth.

Number of CCSR in Ground Plane	Resonance Frequency (GHz)	Notch Depth (dB)	B.W at 3dB (GHz)
One	2.32	−16.54	0.36
Two	2.32	−30.10	0.88
Three	2.32	−46.54	1.01
Four	2.32	−77.93	0.96

3. Sensitivity Analysis

The basic operating principal of microwave sensors is to measure the change in resonance frequency due to interaction with the material under test (MUT). According to the theoretical development, the shift in resonance frequency of sensor due to interaction with the MUT can be expressed as [14]:

$$\frac{\Delta f_r}{f_r} = \frac{\int_v (\Delta \varepsilon E_1 \cdot E_0 + \Delta \mu H_1 \cdot H_0)dv}{\int_v (\varepsilon_0 |E_0|^2 + \mu_0 |H_0|^2)dv} \tag{9}$$

where Δf_r, $\Delta \varepsilon$, $\Delta \mu$ are change in resonance frequency, permittivity, and permeability, respectively. v is the perturbed volume. E_0 and E_1 are the electric fields distribution without and with perturbation, respectively. H_0 and H_1 are the magnetic fields distribution without and with perturbation, respectively. For sensitivity analysis, we selected the sensor based on four CCSR structures due to its sharp transmission curve. MUT is placed in the ground plane without air gap as shown in Figure 8. Sensitivity analysis is performed on permittivity perturbation of MUTs. Four MUTs (teflon, quartz, FR4, and silicon nitrate) with constant dimension (6 mm × 27 mm × 1 mm) are placed under the ground plane and their impact on transmission coefficient of the sensor is shown in Figure 9. Simulated resonance frequencies of the sensors due to interaction with air, teflon, quartz, FR4, and silicon nitrate are 2.32 GHz, 2.17 GHz, 2.04 GHz, 2.0 GHz, and 1.84 GHz, respectively. From Figure 9, it is clear that the relative permittivity of the MUT is inversely proportional to the resonance frequency of the sensor. Permittivity of MUT basically changes the total capacitance of the device, and resonance frequency due to interaction with MUT can be calculated as follows [15]:

$$f = \frac{1}{2\pi\sqrt{L(C_{substrate} + C_{MUT})}} \tag{10}$$

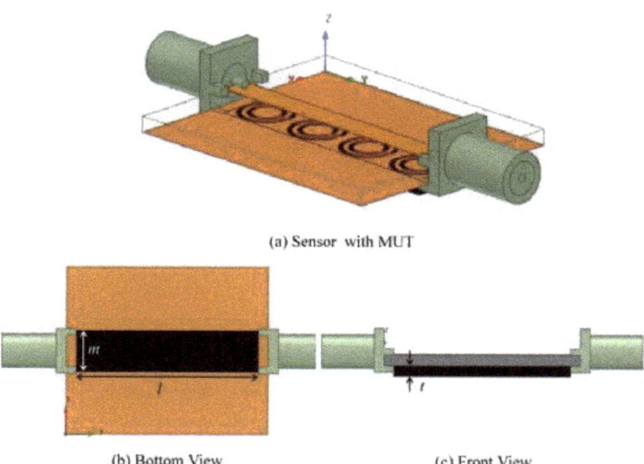

Figure 8. (**a**) Sensor based on four CCSR with materials under test (MUT) in the ground plane without air gap. (**b**) Bottom view of the sensor with MUT (l = 27 mm, and m = 6 mm). (**c**) Front view of the sensor with MUT (t = 1 mm).

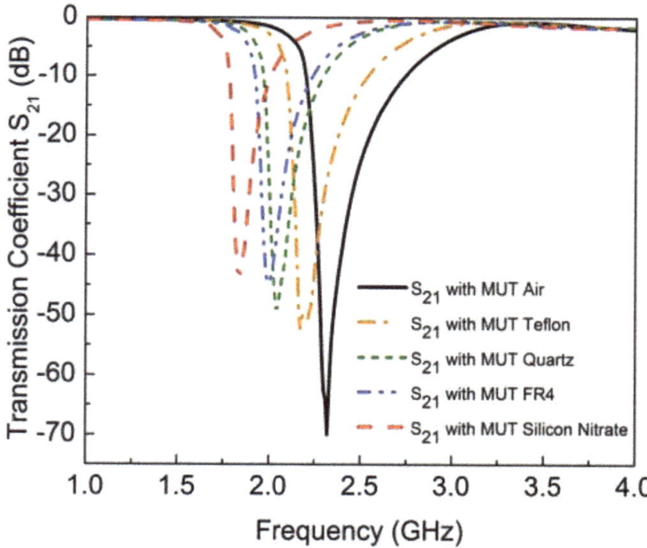

Figure 9. Simulated transmission coefficient S_{21} (dB) of sensor due to interaction with different MUTs. Resonance frequencies of sensors due to interaction with air, teflon, quartz, FR4, and silicon nitrate are 2.32 GHz, 2.17 GHz, 2.04 GHz, 2.0 GHz, and 1.84 GHz, respectively.

4. Fabrication and Measurement

The sensor based on four CCSR structures is fabricated on FR4 substrate using standard PCB fabrication technique and the fabricated prototype is shown in Figure 10b. Vector network analyzer (AV3672series) is used for measurement of transmission coefficient with a frequency sweep of 1 to 4 GHz as shown in Figure 10a. The fabricated sensor is used to measure the resonance frequencies of sensors due to interaction with air, teflon, quartz, FR4, and silicon nitrate as shown in Figure 11. Measured resonance frequencies of the sensors due to interaction with air, teflon, quartz, FR4, and silicon nitrate are 2.29 GHz, 2.20 GHz, 2.06 GHz, 2.02 GHz, and 1.82 GHz, respectively. For comparison, simulated and measured results for the sensor are tabulated in Table 3. The differences between simulated and measured results are very small and can be attributed to fabrication tolerance, conductor, dielectric and radiation losses.

Table 3. Simulated and measured results of sensor for different MUTs.

Material Under Test (MUT)	Relative Permittivity of MUT	Simulated Resonance Frequency (GHz)	Measured Resonance Frequency (GHz)	Difference Between Simulation and Measurement
Air	1	2.32	2.29	0.05
Teflon	2.1	2.17	2.20	0.03
Quartz Glass	3.78	2.04	2.06	0.02
FR4 Epoxy	4.4	2.0	2.02	0.02
Silicon Nitrate	7	1.84	1.82	0.02

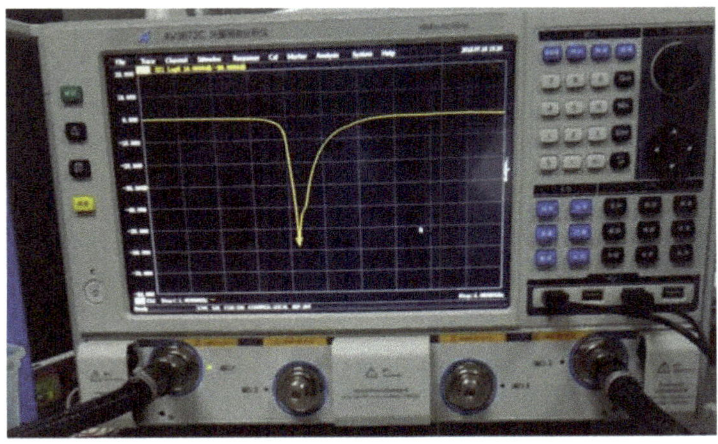

(a)

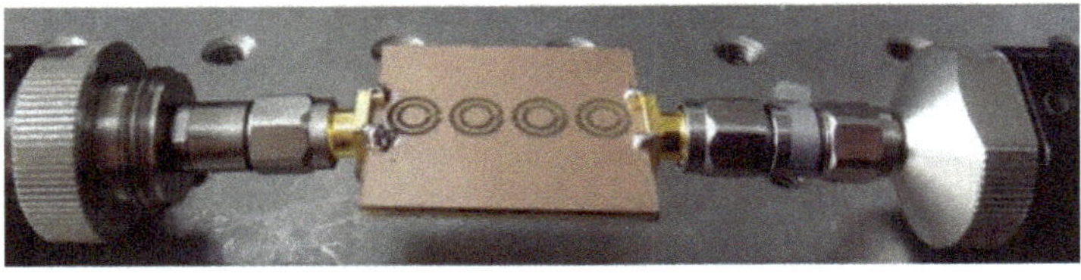

(b)

Figure 10. (a) Photograph of Vector Network Analyzer (VNA) for measurement of sensor. (b) Fabricated prototype of the sensor based on four CCSR.

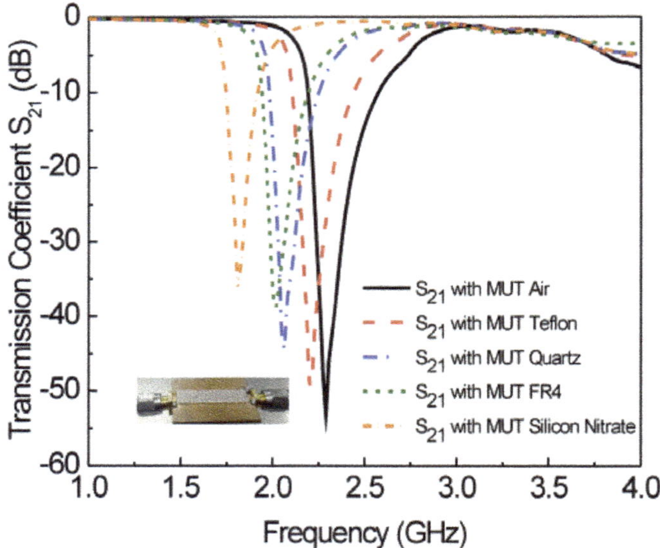

Figure 11. Measured transmission coefficient S_{21} (dB) of sensor due to interaction with different MUTs. Resonance frequencies of sensors due to interaction with air, teflon, quartz, FR4, and silicon nitrate are 2.29 GHz, 2.20 GHz, 2.06 GHz, 2.02 GHz, and 1.82 GHz, respectively.

5. Formulation

The transcendental equation for the proposed sensor is formulated with fitting parameters using measured results. As shown in Figure 11, the resonance frequency of the sensor varies with relative permittivity of MUT. This variation in resonance frequency can be expressed by the following equation [20]:

$$f_{r,MUT} = f_{r,Air} \sqrt{\frac{\varepsilon_{eff,AIR}}{\varepsilon_{eff,MUT}}} \tag{11}$$

where $f_{r,MUT}$ and $f_{r,AIR}$ are resonance frequencies of the sensor with and without MUT, respectively. While $\varepsilon_{eff,MUT}$ and $\varepsilon_{eff,AIR}$ are effective permittivity of MUT and air, respectively. Figure 12 shows the relationship between relative permittivity of MUT and the resonance frequency of sensor due to interaction with MUT. This relationship shows that resonance frequency of the sensor is decreasing by increasing the relative permittivity of MUT. In Ref. [20], resonance frequencies are approximated with a parabolic equation with respect to relative permittivity of MUT. The parabolic equation is given as:

$$f_{r,MUT} = A_1 + A_2 \varepsilon_r' + A_3 \varepsilon_r'^2 \tag{12}$$

where ε_r' is relative permittivity of MUT. A_1, A_2, and A_3 are constant values of polynomial. The reference MUT is air, which has dielectric constant 1. For reference, MUT resonance frequency must be equal to A_1, so Equation (12) can be expanded with respect to $(\varepsilon_r' - 1)$,

$$f_{r,MUT} = A_1 + A_2(\varepsilon_r' - 1) + A_3(\varepsilon_r' - 1)^2 \tag{13}$$

By measuring the results of standard materials (Air, Teflon, and FR4) for which dielectric constants are well known and curve fitting, the constant parameters of Equation (13) are extracted. Finally, the equation becomes:

$$f_{r,MUT} = 2.29 - 0.08297(\varepsilon_r' - 1) + 0.00105(\varepsilon_r' - 1)^2 \tag{14}$$

Equation (14) can be used to predict resonance frequency of known MUTs with permittivity ranges from 1 to 10. To check validity of Equation (14), various MUTs are placed on the sensor and resonance frequencies are extracted through Electromagnetic (EM) simulation. In Figure 13, the solid line shows the resonance frequencies extracted from Equation (14) and dash line shows the resonance frequencies extracted through simulation. It is clear that Equation (14) is fairly reliable for the prediction of resonance frequency of the sensor due to the interaction with MUTs of permittivity ranges up to 10. To calculate the relative permittivity of unknown MUT, the transcendental equation can be expressed as:

$$\varepsilon_r' = \frac{0.08297 - \sqrt{0.00688 - 0.0042\left(2.29 - f_{r,MUT}\right)}}{0.0021} + 1 \tag{15}$$

Equation (15) can be used to calculate relative permittivity of unknown MUTs. In order to check the validity of Equation (15), measured resonance frequencies are used to calculate the relative permittivity of different MUTs and are tabulated in Table 4. It is clear that the relative permittivity values obtained from Equation (15) are very close to the actual values.

Table 4. Relative permittivity evaluation using measured result.

Material Under Test (MUT)	Relative Permittivity (ε_r')	Equation (15) Calculation or ε_r'
Air	1	1.01
Teflon	2.1	2.11
Quartz Glass	3.78	3.88
FR4 Epoxy	4.4	4.41
Silicon Nitrate	7	7.15

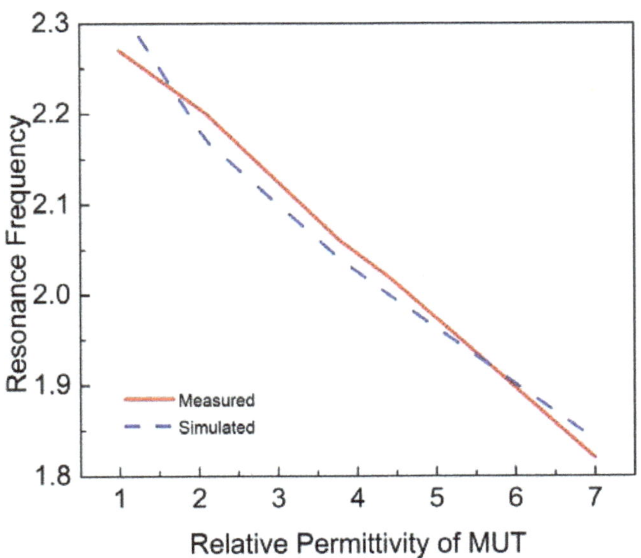

Figure 12. Relative permittivity of MUT versus resonance frequency of the sensor obtained by Electromagnetic (EM) simulation and VNA measurement. Relative permittivity of MUT is inversely proportional to the resonance frequency of sensor.

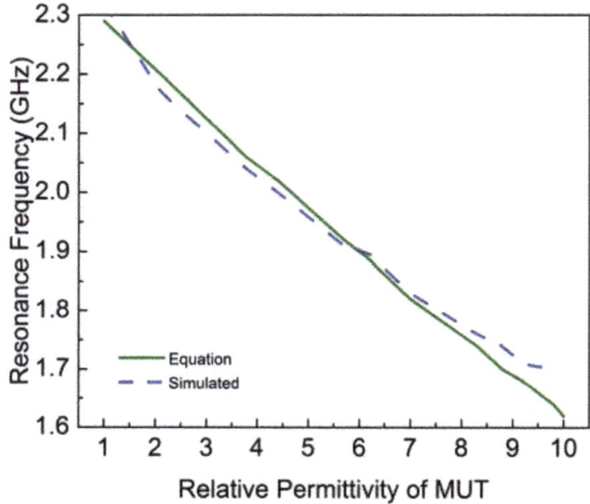

Figure 13. Relative permittivity of MUT versus resonance frequency of the sensor obtained by EM simulation and Equation (14).

6. Conclusions

A complementary metamaterial sensor based on the microstrip transmission line and complementary circular spiral resonator (CCSR) was designed numerically and verified experimentally for nondestructive evaluation of dielectric substrates. Constitutive parameters for the proposed sensor were calculated using ANSYS electronics desktop. A sharp transmission curve was achieved using four CCSR structures in the ground plane. Sensitivity analysis was performed using permittivity perturbation of material under test (MUT). The shift in the resonance frequency of the sensor due to interaction with MUT is presented as a function of permittivity of MUT. The transcendental equation was derived for the sensor to predict the resonance frequency of dielectric materials. Simulated, measured, and formulated results are very close to each other within permittivity range from 1 to 10. The proposed design is very compact and fabrication is easy and inexpensive. Although the main emphasis of this work is for the evaluation of dielectric substrates, in the future the design will be improved for biosensing and security applications.

Author Contributions: T.u.H. proposed the structure of the sensor, and simulated and fabricated it; used the device to test different dielectric substrates and the method to achieve it; participated in the experiments and recording of the data; cataloged references and analyzed the measured results; and prepared the manuscript. C.R. is the research head of the experiment, proposed the design and experimental idea, analyzed the data, and participated in the revision of the paper. X.Z and S.U. participated in the experiment and recorded the data, and participated in the preparation of the manuscript.

Funding: This research was funded by National Natural Science Foundation of China, grant number 61831001, the High-Level Talent Introduction Project of Beihang University, grant number ZG216S1878 and the Youth-Top-Talent Support Project of Beihang University, grant number ZG226S1821.

Conflicts of Interest: The authors declare no conflict of interest.

References

1.	Tiwari, N.K.; Singh, S.P.; Akhtar, M.J. Novel Improved sensitivity planar microwave probe for adulteration detection in edible oils. *IEEE Microw. Wirel. Components Lett.* **2019**, *29*, 164–166. [CrossRef]
2.	Kumari, R.; Patel, P.N.; Yadav, R. An ENG-inspired microwave sensor and functional technique for label-free detection of aspergillus Niger. *IEEE Sens. J.* **2018**, *18*, 3932–3939. [CrossRef]

3. Trabelsi, S.; Nelson, S.O. Microwave sensing of quality attributes of agricultural and food products. *IEEE Instrum. Meas. Mag.* **2016**, *19*, 36–41. [CrossRef]

4. Abdolrazzaghi, M.; Khan, S.; Daneshmand, M. A Dual-Mode Split-Ring Resonator to Eliminate Relative Humidity Impact. *IEEE Microw. Wirel. Components Lett.* **2018**, *28*, 939–941. [CrossRef]

5. Chuma, E.L.; Iano, Y.; Fontgalland, G.; Bravo Roger, L.L. Microwave sensor for liquid dielectric characterization based on metamaterial complementary split ring resonator. *IEEE Sens. J.* **2018**, *18*, 9978–9983. [CrossRef]

6. Falcone, F.; Lopetegi, T.; Baena, J. Effective negative epsilon stopband microstrip lines based on complementary split ring resonators. *IEEE Microw. Wirel. Components Lett.* **2004**, *14*, 280–282. [CrossRef]

7. Pendry, J.B.; Holden, A.J.; Robbins, D.J.; Stewart, W.J. Magnetism from conductors and enhanced nonlinear phenomena. *IEEE Microw. Wirel. Components Lett.* **1999**, *47*, 2075–2084. [CrossRef]

8. Baena, J.D.; Bonache, J.; Martín, F.; Sillero, R.M.; Falcone, F.; Lopetegi, T.; Laso, M.A.G.; García-García, J.; Gil, I.; Portillo, M.F.; et al. Equivalent-circuit models for split-ring resonators and complementary split-ring resonators coupled to planar transmission lines. *IEEE Trans. Microw. Theory Tech.* **2005**, *53*, 1451–1460. [CrossRef]

9. Zarifi, M.H.; Deif, S.; Abdolrazzaghi, M.; Chen, B.; Ramsawak, D.; Amyotte, M.; Vahabisani, N.; Hashisho, Z.; Chen, W.; Daneshmand, M. A microwave ring resonator sensor for early detection of breaches in pipeline coatings. *IEEE Trans. Ind. Electron.* **2017**, *65*, 1626–1635. [CrossRef]

10. Mohd Bahar, A.A.; Zakaria, Z.; Ab Rashid, S.R.; Isa, A.A.M.; Alahnomi, R.A. High-efficiency microwave planar resonator sensor based on bridge split ring topology. *IEEE Microw. Wirel. Components Lett.* **2017**, *27*, 545–547. [CrossRef]

11. Puentes, M.; Maasch, M.; Schubler, M.; Jakoby, R. Frequency multiplexed 2-dimensional sensor array based on split-ring resonators for organic tissue analysis. *IEEE Trans. Microw. Theory Tech.* **2012**, *60*, 1720–1727. [CrossRef]

12. Boybay, M.S.; Ramahi, O.M. Material characterization using complementary split-ring resonators. *IEEE Trans. Instrum. Meas.* **2012**, *61*, 3039–3046. [CrossRef]

13. Haq, T.U.; Ruan, C.; Ullah, S.; Kosar, A. Reconfigurable Ultra Wide Band Notch Filter based on Complementary Metamaterial. In Proceedings of the 2018 IEEE Asia-Pacific Conference on Antennas and Propagation (APCAP), Auckland, New Zealand, 5–8 August 2018.

14. Lee, C.S.; Yang, C.L. Thickness and permittivity measurement in multi-layered dielectric structures using complementary split-ring resonators. *IEEE Sens. J.* **2014**, *14*, 695–700. [CrossRef]

15. Lee, C.S.; Yang, C.L. Complementary split-ring resonators for measuring dielectric constants and loss tangents. *IEEE Microw. Wirel. Components Lett.* **2014**, *24*, 563–565. [CrossRef]

16. Albishi, A.M.; Ramahi, O.M. Highly sensitive microwaves sensors for fluid concentration measurements. *IEEE Microw. Wirel. Components Lett.* **2018**, *28*, 287–289. [CrossRef]

17. Zhang, X.; Ruan, C.; Haq, T.; Chen, K. High-Sensitivity Microwave Sensor for Liquid Characterization Using a Complementary Circular Spiral Resonator. *Sensors* **2019**, *19*, 787. [CrossRef] [PubMed]

18. HONG, J.-S. *Microstrip filters for RF/Microwave Applications*; Wiley & Sons: New York, NY, USA, 2011; ISBN 9780470408773.

19. Chen, X.; Grzegorczyk, T.M.; Wu, B.I.; Pacheco, J.; Kong, J.A. Robust method to retrieve the constitutive effective parameters of metamaterials. *Phys. Rev. E Stat. Physics Plasmas Fluids Relat. Interdiscip. Top.* **2004**, *70*, 7. [CrossRef]

20. Lim, S.; Kim, C.Y.; Hong, S. Simultaneous Measurement of Thickness and Permittivity by Means of the Resonant Frequency Fitting of a Microstrip Line Ring Resonator. *IEEE Microw. Wirel. Components Lett.* **2018**, *28*, 539–541. [CrossRef]

Article

Low Concentration Response Hydrogen Sensors Based on Wheatstone Bridge

Hongchuan Jiang [1,*], Xiaoyu Tian [1] , Xinwu Deng [1], Xiaohui Zhao [1] , Luying Zhang [1], Wanli Zhang [1], Jianfeng Zhang [2] and Yifan Huang [2]

[1] State Key Laboratory of Electronic Thin Films and Integrated Devices, University of Electronic Science and Technology of China, Chengdu 610054, China; uestctxy@163.com (X.T.); xwdeng@uestc.edu.cn (X.D.); xhzhao@uestc.edu.cn (X.Z.); luyingz@163.com (L.Z.); wlzhang@uestc.edu.cn (W.Z.)

[2] National Key Laboratory of Science and Technology on Vacuum Technology and Physics, Lanzhou Institute of Physics, Lanzhou 730000, China; zhangjianfeng510@spacechina.com (J.Z.); huangyifan@spacechina.com (Y.H.)

* Correspondence: hcjiang@uestc.edu.cn; Tel.: +86-28-6183-1181

Received: 24 January 2019; Accepted: 1 March 2019; Published: 4 March 2019

Abstract: The PdNi film hydrogen sensors with Wheatstone bridge structure were designed and fabricated with the micro-electro-mechanical system (MEMS) technology. The integrated sensors consisted of four PdNi alloy film resistors. The internal two were shielded with silicon nitride film and used as reference resistors, while the others were used for hydrogen sensing. The PdNi alloy films and SiN films were deposited by magnetron sputtering. The morphology and microstructure of the PdNi films were characterized with X-ray diffraction (XRD). For efficient data acquisition, the output signal was converted from resistance to voltage. Hydrogen (H_2) sensing properties of PdNi film hydrogen sensors with Wheatstone bridge structure were investigated under different temperatures (30 °C, 50 °C and 70 °C) and H_2 concentrations (from 10 ppm to 0.4%). The hydrogen sensor demonstrated distinct response at different hydrogen concentrations and high repeatability in cycle testing under 0.4% H_2 concentration. Towards 10 ppm hydrogen, the PdNi film hydrogen sensor had evident and collectable output voltage of 600 μV.

Keywords: hydrogen sensors; PdNi thin films; Wheatstone bridge; low concentration

---

1. Introduction

Hydrogen (H_2) is one of the most potential and cleanest energy sources [1]. Considering its low minimum ignition energy and wide flammable range (4–75%), a responsive sensor is essential to detect hydrogen concentration [1–3]. Fast and accurate detection of hydrogen concentration is vital to prevent hydrogen leakage when using liquid hydrogen and other aerospace operations in space [2]. Based on different working principles, the hydrogen sensor can be divided into electrical, electrochemical, optical type, work function based, etc. Among those sensors, work function based sensors typically need to work at elevated temperature in order to maintain high sensitivity [2,3]. While resistance based sensors are extensively applied due to ambient working temperature and convenience for measurement [2,4–6].

In the past decades, palladium (Pd) and palladium alloy have been widely used as the sensitive materials for metallic resistor type hydrogen sensor due to its high solubility for hydrogen [2,4,7,8]. Although pure Pd could deliver a high output, fast response and superior selectivity of hydrogen, there are still some shortcomings, such as structural deformations and the consequent hysteretic resistance behavior [2,4]. In order to solve these problems, nanostructured materials and Pd alloy have been used as hydrogen sensitive materials [9–15]. Ozturk et al. reported that when exposed to 10% hydrogen, the 6 nm Pd films deposited on flexible substrates exhibited good reversibility and good response [3]. Lee et al. reported that hydrogen absorption and desorption did not affect the macroscopic structural deformation of PdNi. The sensitivity decreased linearly with increasing Ni content in PdNi film [14]. Hoffheins et al. reported a sensor consisting of four pure resistors with Wheatstone bridge structure; two of them were shielded with the borosilicate-based glass as hydrogen permeation barrier and served as reference resistors compensating for changes in the resistance of the palladium due to temperature variations. The thick film sensor could detect hydrogen concentration at temperatures from 0 to 200 °C. However, the low concentration limit for hydrogen sensing was no less than 0.5% [16]. Moreover, the fabrication process of the sensor was not applicable for mass production.

In this study, a PdNi film hydrogen sensor with Wheatstone Bridge structure was designed and fabricated with micro-electro-mechanical system (MEMS) technology. The integrated sensors consisted of four PdNi alloy film resistors: two of them were shielded with silicon nitride film and used as reference resistance, while the others were used for hydrogen sensing. The output resistance signal was converted to millivolt output voltage signal for easy data acquisition. In addition, the property of the sensor was ameliorated by annealing treatment. Additionally, the performances of the Wheatstone bridge thin film hydrogen sensor were characterized and discussed.

2. Materials and Methods

2.1. Design and Fabrication of the Sensor

The schematic of the hydrogen sensor is shown in Figure 1. The sample was designed within the size of 3 cm × 3.25 cm × 0.43 mm. Single-sided polished silicon wafers were selected as the substrates. PdNi alloy film was used as the hydrogen sensing resistor. The bond pads of the PdNi were covered with Au film for precise resistance measurement. Additioally, line width of the PdNi line film resistance was 100 μm. The resistance of all of the resistors was designed as 8.4 kΩ.

Figure 1. Schematic of the hydrogen sensors.

As shown in the Figure 2, a Wheatstone bridge type hydrogen sensor was made up of four PdNi film resistors. An external voltage source U_{in} was applied to Wheatstone bridge type hydrogen sensor. The output voltage (U_{out}) can be expressed by:

$$U_{out} = \left(\frac{R_1 + \Delta R_1}{R_1 + R_2 + \Delta R_1} - \frac{R_3}{R_3 + R_4 + \Delta R_4} \right) U_{in} - U_0 \tag{1}$$

when the resistance of four resistors is the same ($R = R_1 = R_2 = R_3 = R_4$), it can be simplified as:

$$U_{out} = \frac{\Delta R}{2R + \Delta R} U_{in} - U_0 = \frac{\Delta R}{2R + \Delta R} U_{in} \tag{2}$$

where U_{in} and U_0 are the voltage of external voltage source and initial bias of the Wheatstone bridge, respectively. Thus, the output voltage U_{out} is directly related with ΔR. The output resistance signal was converted to millivolt output voltage signal for easy data acquisition.

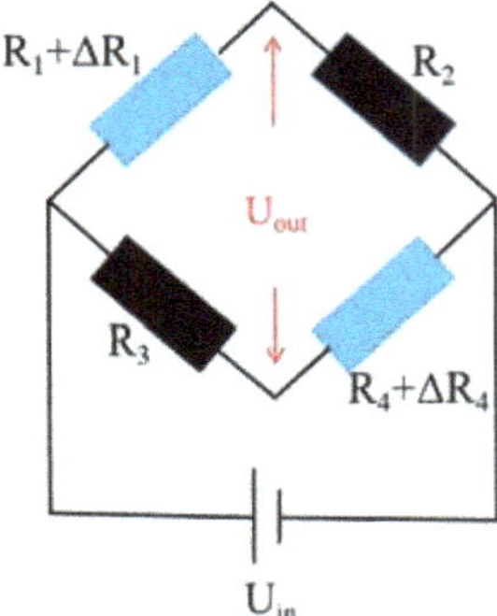

Figure 2. Structure diagram of Wheatstone bridge type hydrogen sensor.

The fabrication process of the hydrogen sensor was as follows: prior to the film deposition, the substrate was ultrasonic-cleaned successively with acetone, alcohol, and deionized water for 10 min. After that, the substrate was blow-dried with nitrogen. Next, PRI-4000A photoresist was spin-coated onto the substrate and patterned with lithographic technology. Afterwards, the PdNi film with the thickness of 100 nm was deposited by direct current magnetron sputtering and patterned with lift-off technique. Then, the silicon nitride layer with the thickness of 160 nm was deposited by RF magnetron sputtering and patterned on the two reference resistors. At last, Au film with the thickness of 220 nm was deposited by DC magnetron sputtering and patterned on the bond pads of PdNi resistors. Sputtering parameters of all the deposition process are included in Table 1. The crystal structure of the PdNi films was examined with X-ray diffraction (XRD, D/MAX-rA diffractometer, Rigaku, Japan) with a scan range from 35° to 50°. The surface morphology of the PdNi film was examined by scanning electron microscopy (SEM, Inspect F50, FEI, United States of America).

Table 1. Sputtering parameters of different films.

Material	Base Pressure (Pa)	Sputtering Pressure (Pa)	Sputtering Power (W)	Temperature (°C)
PdNi	8×10^{-4}	0.3	60	RT
Si$_3$N$_4$	8×10^{-4}	0.5	200	RT
Au	8×10^{-4}	0.3	60	RT

After the fabrication, the sample was vacuum annealed in nitrogen for 2 h at 300 °C. The sample was sliced into a single sensor with a size of 3 cm × 3.25 cm and gold wire was ball welded on the bond pad for data collection. The four resistors were connected accordingly with the external circuit to form the Wheatstone bridge. The photo of the fabricated hydrogen sensor is shown in Figure 3.

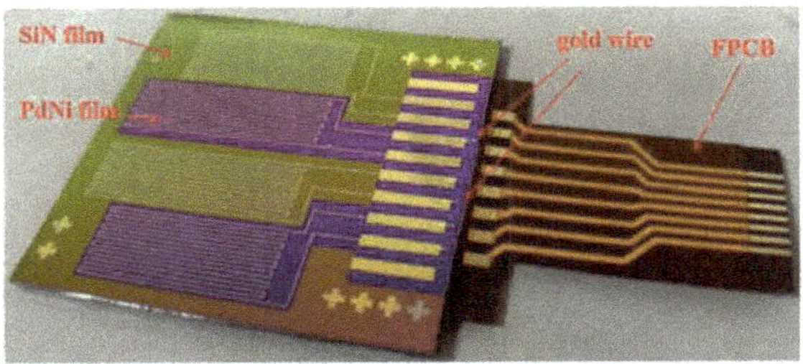

Figure 3. Photo of the fabricated hydrogen sensor.

2.2. Hydrogen Sensing Tests

The hydrogen measuring system consisted of mass flow controllers (MFC), gas mix chamber, gas test chamber, temperature-controlled chamber, Keithley 2182 Nanovoltmeter and the power source (GPC-60300, Gwinstek, China), as shown in Figure 4. Before the measurement, the gas chamber was continuously purged with pure nitrogen for 2 h with the flow rate of 100 sccm (standard cubic centimeters per minute). Next, nitrogen and hydrogen gas with different ratios was transmitted to the mix chamber. After mixing, the gas mixture was delivered to the gas chamber. The flow rate the gas mixture was controlled to be 100 sccm. The output voltage of the Wheatstone Bridge was acquired with a LabVIEW program (National Instruments) under constant voltage mode with a source voltage of 10 V.

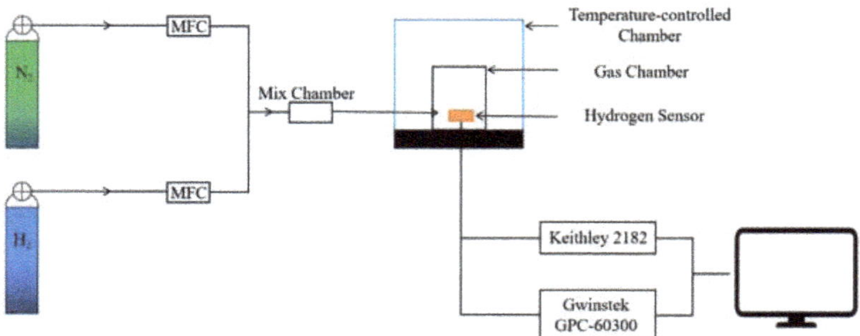

Figure 4. Schematic of the hydrogen measuring system.

3. Results and Discussions

XRD pattern of the PdNi thin film was shown in Figure 5, two diffraction peaks localized at 40.64° and 47.69° reflect face-centered cubic (fcc) crystal structures of Pd and can be indexed as (111) and (200) planes, respectively. The peak positions are gradually shifted to higher angle with increasing Ni concentration in the PdNi films [11]. The full width half maximum (FWHM) of the peaks decreased after annealing, indicating the increment of the crystal size calculated with the Scherrer Equation [7].

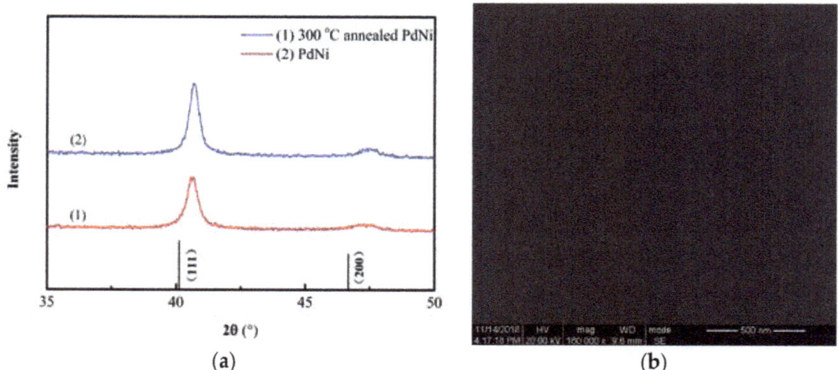

Figure 5. (a) X-ray diffraction (XRD) patterns of (1) PdNi and (2) 300 °C annealed PdNi film; (b) SEM image of PdNi thin film.

The repeatability of the hydrogen sensor towards 0.4% hydrogen at different temperatures (30 °C, 50 °C and 70 °C) was revealed in the Figure 6. As the test temperature increased, the stability and zero drift of the sensor were partially improved despite the attenuation of the output signal. This phenomenon can be explained with the Sievert' law and Sievert constant (K) which is defined as the solubility of gas molecules in metal materials:

$$\ln K = -\frac{\Delta_s H}{RT} + \frac{\Delta_s S}{R} \tag{3}$$

where $\Delta_s H$ is the molar enthalpy of solution, $\Delta_s S$ the molar entropy of solution, R the gas constant and T the environment temperature [3,17,18]. Equation (3) explains clearly that there was an inverse relationship between the temperature and sensitivity of the metal film resistance sensor [3]. The solubility of H in PdNi alloy films decreased with increased test temperature, which results in reducing maximum response of Wheatstone bridge type hydrogen sensor.

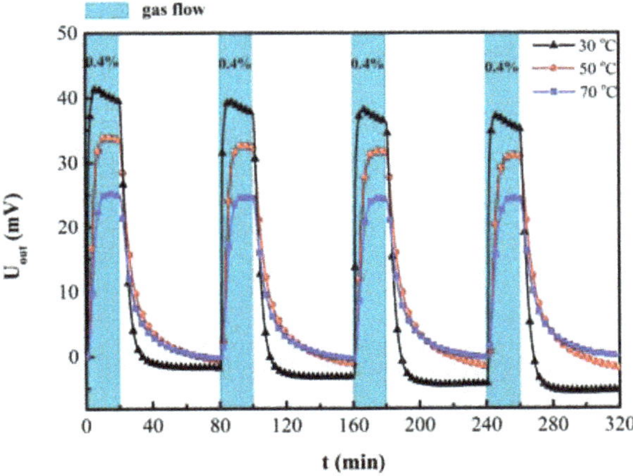

Figure 6. Repeatability of the hydrogen sensor at 30 °C, 50 °C and 70 °C under 0.4% H_2.

The response time and recovery time were defined as the consumed time to reach 90% of the final stable value [12]. When exposed to hydrogen, the output voltage of the sensor increased rapidly. After switching to pure nitrogen, the output voltage of the sensor decreased sharply, as shown in Figures 6 and 7a. When the hydrogen concentration increased from 200 ppm to 800 ppm, the output voltage of (2) the unannealed hydrogen sensor increased from 10.225 mV to 15.434 mV; the response time and recovery time also decreased gradually, as shown in Figure 7a. These results indicate that with increasing hydrogen concentration, the diffusion rate of hydrogen and the rate of hydride formation increased [2,4,5,17]. With a same test time, the output value of unannealed sensor did not attain the equilibrium at 200 ppm. Compared with unannealed sample, the output response of the annealed sensor was somewhat reduced; however, its zero drift and repeatability were greatly improved. It has been reported that annealing treatment could alter the microstructure of Pd alloy film into influence H absorption characteristics. Hydrogen solubilities in dilute phase decrease with enhancing alloy homogeneity [19]. After annealing at 300 °C, the PdNi peak is apparently stronger due to the grain size growth in the Figure 5a. Due to the low annealing temperature, the microstructure of the film in the SEM image might not be significantly different. Grain size growth might be one of the factors causing change of H absorption. In addition, the internal stress in alloy film was released after annealing treatment, which might be another factor for the fast response time [8,20]. The output signal, the response and recovery time of the annealed hydrogen sensor all decrease after annealing treatment.

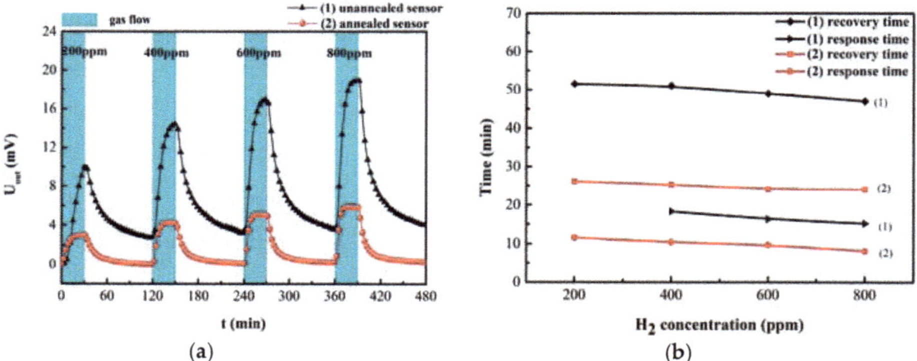

Figure 7. (a) Response curve of (1) unannealed sensor and (2) annealed hydrogen sensor at 50 °C under different H_2 concentrations; (b) the response time and recovery time of (1) and (2).

Figure 8 displays the response of the annealed hydrogen sensor at low H_2 concentration. The annealed sensor still has a detectable output voltage of 600 μV at 10 ppm hydrogen. In addition, compared with single PdNi film resistance hydrogen sensor [12], the response time and recovery time of the Wheatstone bridge sensor were significantly reduced and the zero drift was greatly ameliorated. It should be noted that in sharp contrast to previous report [14–16], this sensor still had an obvious output response under 10 ppm hydrogen, which makes it applicable for wide range hydrogen concentration detection (10 ppm-0.4%). Furthermore, the sensor was fabricated on silicon substrates with MEMS technology, which makes it much easier for mass production and integration [21].

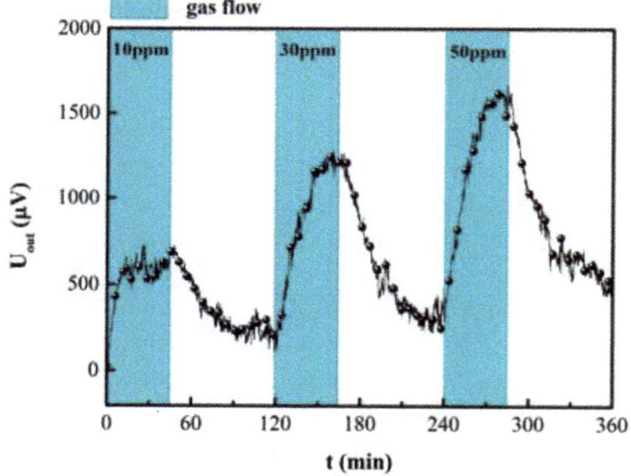

Figure 8. Response curve of the annealed hydrogen sensor at 50 °C.

4. Conclusions

The highly sensitive Wheatstone bridge type hydrogen sensor based on PdNi alloy thin films was fabricated and tested. The output resistance signal was converted to millivolt output voltage signal for easy data acquisition. Compared with unannealed sample, the annealed sensor had excellent output response, faster response time and recovery time. Furthermore, the sensor demonstrated obvious output response with excellent stability and reliability over wide range of hydrogen concentration (10 ppm-0.4%). This sensor was fabricated on silicon substrates with MEMS technology, which makes it much easier for mass production and integration.

Author Contributions: This work presented in this paper was a collaboration of all the authors. H.J. and W.Z. conceived and designed the experiments; X.T., L.Z. and X.D. performed the experiments; H.J., X.T. and X.Z. analyzed the data and wrote the paper; J.Z. and Y.H. revised the paper.

Funding: This work was supported by the foundation of the National Key Research and Development Program of China (No. 2016YFB0501303) and the National Key Laboratory of Science and Technology on Vacuum Technology and Physics (No. ZWK1701).

Conflicts of Interest: The authors declare no conflict of interest.

References

1.   Schlapbach, L.; Zuttel, A. Hydrogen-storage materials for mobile applications. *Nature* **2001**, *414*, 353–358. [CrossRef] [PubMed]

2.   Hübert, T.; Boon-Brett, L.; Black, G.; Banach, U. Hydrogen sensors—A review. *Sens. Actuators B Chem.* **2011**, *157*, 329–352. [CrossRef]

3.   Ozturk, S.; Kilinc, N. Pd thin films on flexible substrate for hydrogen sensor. *J. Alloy. Compd.* **2016**, *674*, 179–184. [CrossRef]

4.   Paglieri, S.N.; Way, J.D. Innovations in palladium membrane research. *Sep. Purif. Methods* **2002**, *31*, 1–169. [CrossRef]

5.   Sakamoto, Y.; Takashima, I. Hysteresis behaviour of electrical resistance of the Pd-H system measured by a gas-phase method. *J. Phys. Condensed. Matter.* **1996**, *8*, 10511–10520. [CrossRef]

6.   Wang, B.; Zhu, Y.; Chen, Y.; Song, H.; Huang, P.; Dao, D.V. Hydrogen sensor based on palladium-yttrium alloy nanosheet. *Mater. Chem. Phys.* **2017**, *194*, 231–235. [CrossRef]

7.   Monshi, A.; Foroughi, M.R.; Monshi, M.R. Modified Scherrer Equation to Estimate More Accurately Nano-Crystallite Size Using XRD. *World J. Nano Sci. Eng.* **2012**, *2*, 154–160. [CrossRef]

8.   Zhao, Z.; Carpenter, M.A. Annealing enhanced hydrogen absorption in nanocrystalline Pd/Au sensing films. *J. Appl. Phys.* **2005**, *97*, 124301. [CrossRef]

9.   Yoshimura, K.; Nakano, S.; Uchinashi, S.; Yamaura, S.; Kimura, H.; Inoue, A. A hydrogen sensor based on Mg–Pd alloy thin film. *Meas. Sci. Technol.* **2007**, *18*, 3335–3338. [CrossRef]

10.   Phan, D.-T.; Chung, G.-S. Reliability of hydrogen sensing based on bimetallic Ni–Pd/graphene composites. *Int. J. Hydrog. Energ.* **2014**, *39*, 20294–20304. [CrossRef]

11.   Rashid, T.-R.; Phan, D.-T.; Chung, G.-S. A flexible hydrogen sensor based on Pd nanoparticles decorated ZnO nanorods grown on polyimide tape. *Sens. Actuators B: Chem.* **2013**, *185*, 777–784. [CrossRef]

12.   Jiang, H.C.; Huang, M.; Yu, Y.B.; Tian, X.Y.; Zhao, X.H.; Zhang, W.L.; Zhang, J.F.; Huang, Y.F.; Yu, K. Integrated Temperature and Hydrogen Sensors with MEMS Technology. *Sensors* **2017**, *18*, 94. [CrossRef] [PubMed]

13.   Wadell, C.; Nugroho, F.A.A.; Lidstrom, E.; Iandolo, B.; Wagner, J.B.; Langhammer, C. Hysteresis-Free Nanoplasmonic Pd–Au Alloy Hydrogen Sensors. *Nano Lett.* **2015**, *15*, 3563–3570. [CrossRef] [PubMed]

14.   Lee, E.; Lee, J.M.; Lee, E.; Noh, J.; Joe, J.H.; Jung, B.; Lee, W. Hydrogen gas sensing performance of Pd–Ni alloy thin films. *Thin Solid Films* **2010**, *519*, 880–884. [CrossRef]

15.   Chachuli, S.A.M.; Hamidon, M.N.; Mamat, M.S.; Ertugrul, M.; Abdullah, N.H. A Hydrogen Gas Sensor Based on TiO$_2$ Nanoparticles on Alumina Substrate. *Sensors* **2018**, *18*, 2483. [CrossRef] [PubMed]

16.   Hoffheins, B.S.; Maxey, L.C.; Holmes, W., Jr.; Lauf, R.J.; Salter, C.; Walker, D. Development of low cost sensors for hydrogen safety application. In Proceedings of the 10th Annual National Hydrogen Association Meeting, Vienna, Austria, 7–9 April 1999. NREL/CP-570-269.

17.   Jewell, L.L.; Davis, B.H. Review of Absorption and Adsorption in the Hydrogen–Palladium System. *Appl. Catal. A Gen.* **2006**, *310*, 1–15. [CrossRef]

18.   Wicke, E.; Brodowsky, H.; Zuchner, H. Hydrogen in Palladium and Palladium Alloys. *Met. Finish.* **1996**, *95*, 73–155.

19.   Noh, H.; Luo, W.F.; Flanagan, T.B. The Effect of Annealing Pretreatment of Pd-Rh Alloys on Their Hydrogen Solubilities and Thermodynamic Parameters for H$_2$ Solution. *J. Alloy. Compd.* **1993**, *196*, 7–16. [CrossRef]

Sensors **2019**, *19*, 1096

20. Hao, M.M.; Wu, S.H.; Han, Z.; Ye, W.B.; Wei, X.B.; Wang, X.R.; Zhi, C.; Li, S.B. Room-temperature and fast response hydrogen sensor based on annealed nanoporous palladium film. *J. Mater. Sci.* **2016**, *51*, 2420–2426. [CrossRef]
21. Nabki, F.; Allidina, K.; Ahmad, F.; Cicek, P.V.; El-Gamal, M.N. A Highly Integrated 1.8 Ghz Frequency Synthesizer Based on a Mems Resonator. *IEEE J. Solid-St. Circ.* **2009**, *44*, 2154–2168. [CrossRef]

Article

Low-Hysteresis and Fast Response Time Humidity Sensors Using Suspended Functionalized Carbon Nanotubes

Shivaram Arunachalam *, Ricardo Izquierdo and Frederic Nabki

Department of Electrical Engineering, École de Technologie Supérieure, Montreal, QC H3C 1K3, Canada; ricardo.izquierdo@etsmtl.ca (R.I.); Frederic.Nabki@etsmtl.ca (F.N.)
* Correspondence: shiva-ram.arunachalam.1@ens.etsmtl.ca; Tel.: +1-514-572-2971

Received: 26 December 2018; Accepted: 5 February 2019; Published: 7 February 2019

Abstract: A humidity sensor using suspended carbon nanotubes (CNTs) was fabricated using a low-temperature surface micromachining process. The CNTs were functionalized with carboxylic acid groups that facilitated the interaction of water vapor with the CNTs. The humidity sensor showed a response time of 12 s and a recovery time of 47 s, along with superior hysteresis and stable performance. The hysteresis curve area of the suspended structure is 3.6, a 3.2-fold reduction in comparison to the non-suspended structure. A comparative study between suspended and non-suspended devices highlights the advantages of using a suspended architecture.

Keywords: carbon nanotubes; suspended beam; non-suspended beam; humidity sensing; functionalized

---

1. Introduction

Carbon nanotubes (CNTs), first discovered by Iijima [1], are one of the most studied materials for gas sensing applications [2]. The inherent properties of CNTs, such as high surface area to volume ratio and a hollow structure, along with the ease of surface modification, have led to the use of both single- and multiwalled CNTs to detect a variety of gases with much success [3]. Sensing mechanisms can include change in resistance [4], capacitance [5,6], frequency [7], etc. To date, most of the humidity sensors based on CNTs have been based on functionalized CNTs and have shown considerable promise for commercial usage. Humidity sensors have considerable usage in household, automotive, and medical applications [8]. The performance and stability of sensors are important parameters that are the subject of continuous research using different materials and methods [9]. Fast response and recovery times and low hysteresis of a humidity sensor are important parameters to consider, and recovery performance during prolonged exposure to high humidity levels is of interest for sensors to be commercially deployed. Accordingly, these parameters are considered in this work for the proposed devices.

There are many works in the literature focusing on the fabrication of humidity sensors based on CNTs. Multiwalled CNTs (MWCNTs)-based humidity sensors were fabricated and studied by many research groups (e.g., [5,10,11]). An ultrafast humidity sensor was fabricated by Chan et al. [12] using composite films of polyimide and MWCNTs with a response time of 5 s and a recovery time of 8 min. Liu et al. [13] developed a MWCNTs-based humidity sensor deposited by dielectrophoresis on interdigitated electrodes. Yeow et al. [14] used MWCNTs deposited on a stainless-steel substrate to fabricate a capacitive humidity sensor with a response time of 75 s. An MWCNTs-based humidity sensor was studied using different testing frequencies (AC current) with a response time of 16 s [15]. Researchers have also used single-walled CNTs to detect humidity [16]. However, most of the works mentioned above focus on the hysteresis performance of the proposed sensors. Accordingly, this work aims at characterizing this important parameter as well.

Suspended CNTs have also been used previously to detect gases such as NO_2 [17]. One of the main advantages of using a suspended architecture is the significantly reduced hysteresis in the devices [18]. This, combined with the increased surface area of adsorption is ideal for sensing applications. This has been demonstrated in our previous work [19]. Reduced hysteresis in a sensor leads to increased reliability and measurement consistency, which are two important parameters for sensor performance. However, suspended CNTs are difficult to fabricate and require complex and often expensive tools. There is also the question of yield in traditionally followed methods to obtain suspended CNTs where individual CNTs or a cluster of CNTs are grown between pre-fabricated electrodes using high-temperature chemical or physical vapor deposition processes. This approach is generally used to study intrinsic properties such as transport phenomena [20], phonon interactions [21], etc. Thus, suspended CNTs have rarely been used in sensing applications, even though they offer significant inherent advantages.

In our previous work [19], the advantages of pristine suspended CNTs have been demonstrated. The suspended CNTs showed near-zero hysteresis and better response and recovery times than non-suspended CNTs. Moreover, the fabrication process proposed had a low temperature budget, making the sensors suitable for integration above integrated circuits. However, the response and recovery times were relatively long, curtailing commercial viability. This work is an improvement over the previous work by using functionalized single-walled CNTs. The suspended functionalized CNTs also showed near zero hysteresis, remarkable repeatability along with fast response and recovery times. To further demonstrate the advantages of using a suspended architecture, the device performance is compared to a traditional non-suspended CNT device using the same fabrication process.

2. Materials and Methods

The detailed fabrication process has been discussed previously in [19]. Notably, the fabrication process features a very low temperature budget (i.e., below 110 °C) and is amenable to integration above integrated circuits. The following is a brief overview of the fabrication process.

The process is carried out on a silicon (Si) substrate. SU8, an epoxy-based negative photoresist is used as a sacrificial layer to obtain suspended CNTs. SU8 was chosen as the sacrificial material due to its uniform surface profile and ease of availability. The SU8 is partially crosslinked by UV lithography. The crosslinked SU8, of thickness 3.6 μm, is used as an anchor for the CNT beam, while the uncrosslinked SU8 is used as a sacrificial layer. The uncrosslinked SU8 is prevented from dissolving in further processing steps by depositing a 60-nm barrier layer of aluminum by filament evaporation. The CNTs used in this work are obtained pre-functionalized with a carboxylic acid group (-COOH) from Carbon Solutions Inc. (Riverside, CA, USA) and used without any further material processing. 0.125 g of the as-obtained CNTs in powder form were dispersed in a 1 wt % SDS solution. The as-prepared solution was then ultrasonicated for 6 h at room temperature, resulting in a uniformly dispersed CNT solution. The solution is then ultra-centrifuged at 47,300 rpm for 60 min. The top half of the solution is then decanted for further use. Then, the CNT films are formed using vacuum filtration. Vacuum filtration enables the formation of a homogenous film with a uniform distribution of CNTs. This process also ensures strict control over film thickness. The CNT films used here are 0.7 μm thick. The 20 nm thick aluminum (Al) electrodes are deposited on the CNTs before release to form the suspended CNT beams. The schematic of the resulting suspended CNT beam cross section is shown in Figure 1a and a SEM micrograph of a fabricated beam is shown in Figure 1b.

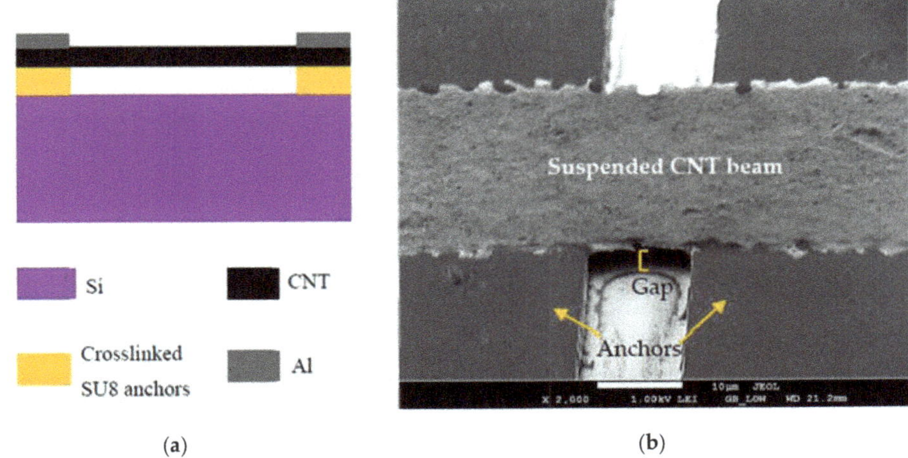

(a)   (b)

Figure 1. (**a**) Schematic of the suspended CNT beam (**b**) SEM micrograph.

3. Results

The device characterization is done in a humidity chamber connected to an external humidifier, which acts as the humidity source. The sensors are interconnected using probes aligned using micro-positioners. The humidity percentage in the chamber was verified using a high-precision sensor embedded within the chamber. The resistance of the devices was measured using a Keithley Digital Multimeter (Cleveland, OH, USA) connected to a computer for data acquisition.

3.1. Humidity Response

To study the humidity sensing properties of the CNT-based sensors, the resistance of the devices was measured as a function of relative humidity. The humidity was measured from a base of 15% RH to a maximum of 98% RH gradually increased in steps of ∼10% RH per 2 min. The results are shown in Figure 2 for both suspended and non-suspended devices. The inset plots of Figure 2a,b show the average hysteresis profile of six measured devices of each type (i.e., suspended and non-suspended) while the outset shows the device response of a typical device of each type. A comparison of the hysteresis profile of a single device to the average profile shows that the suspended devices have a consistent response to humidity. The average base resistance of the suspended CNTs for six measured devices at 15% RH is 1.3 kΩ and it rises to an average of around 3.5 kΩ at 98% RH. In comparison, the non-suspended devices had an average base resistance of 3.3 kΩ at 15% RH and 5.1 kΩ at 98% RH. Moreover, the suspended devices exhibit minimal hysteresis with an average hysteresis curve area of 3.56, even after eight measurement cycles. In comparison, the non-suspended sensors exhibited significant hysteresis, with an average hysteresis curve area of 11.21. This represents a 3.2-fold improvement in the hysteresis behavior of the suspended sensor over the non-suspended sensor. The hysteresis curve area was calculated using the in-built Integrate function of the Origin software package. The lack of hysteresis in the suspended sensor can be attributed to the lack of substrate effects such as charge traps [18], which could alter the pathway of the charge carriers in the nanotube network. The charge traps and grain boundaries within the substrate alter the conductivity and hence the non-suspended sensor also shows higher resistance.

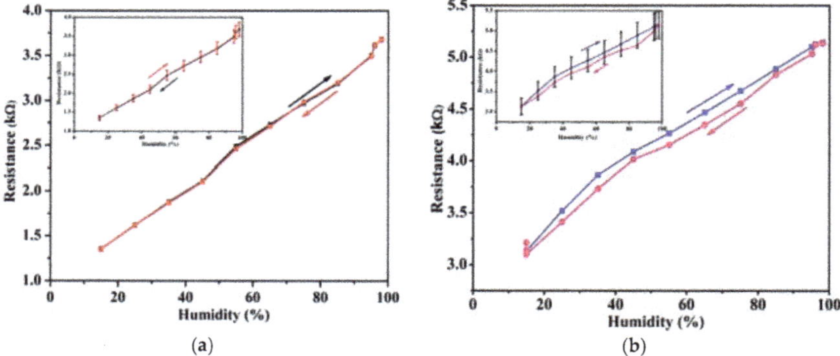

Figure 2. (**a**) Plot of the humidity response of the suspended CNTs, and (**b**) Plot of the humidity response of the non-suspended CNTs. The hysteresis characteristic is also shown. The insets in each figure show the average hysteresis profile of 6 identical devices measured under similar testing conditions.

3.2. Response Time, Recovery Time, and Sensitivity

The response time, defined as the time taken to achieve 10-90% of the total resistance change, is measured to be 12 s for the suspended sensors and around 28 s for the non-suspended sensors, as shown in Figure 3a,c, respectively. The 2.3-fold shorter response time of the suspended sensors is due to the increased surface area for the adsorption of water molecules. Since the film is suspended, the molecules can adhere to both the top and bottom surface of the CNT beam. The recovery time, defined as the time taken by the sensor to go from 90% to 10% of its original resistance value is measured to be 47 s for the suspended sensors and 54 s for the non-suspended sensors, as shown in Figure 3b,d, respectively.

Consistent performance of the sensor is necessary for long-term applications. Thus, repeatability of the sensors was verified by subjecting them to continuous humidity cycles by varying humidity from 15% to 98% RH and recording the resistance at an interval of every 5 s. As seen in Figure 4a, the suspended sensors showed remarkable stability and repeatability. In comparison, the non-suspended sensors showed relatively stable performance with a drift in performance after three humidity cycles, as shown in Figure 4b. This further demonstrates the advantages of using suspended CNTs over traditional non-suspended CNTs. The excellent repeatability is due to the ease of adsorption and desorption of the water molecules from the surface of the suspended CNT beam.

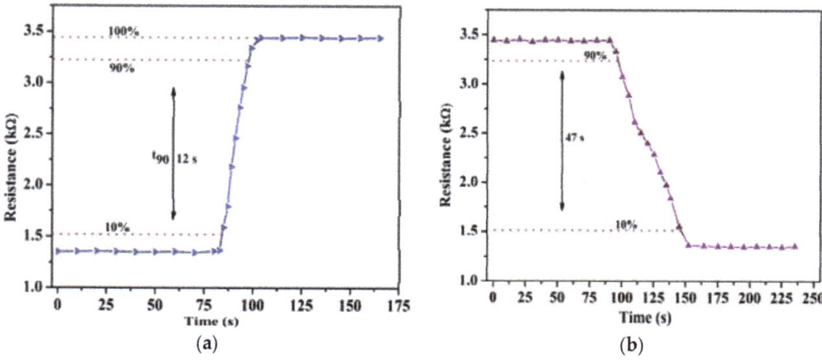

Figure 3. *Cont.*

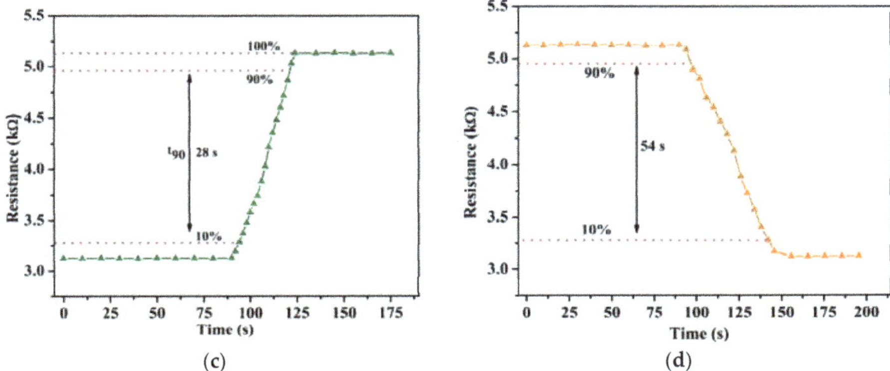

(c)

(d)

Figure 3. (**a**) Response time and (**b**) recovery time of the suspended CNTs; (**c**) response time and (**d**) recovery time of the non-suspended CNTs. The humidity was increased gradually in steps of 10% RH.

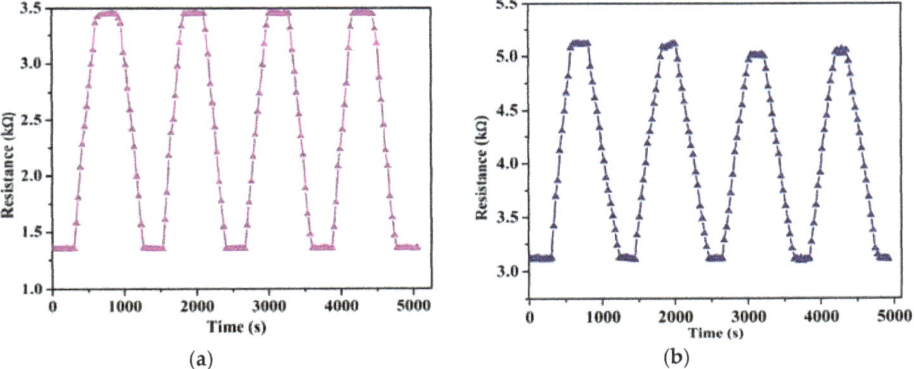

(a)

(b)

Figure 4. (**a**) Repeatability measurement over four humidity cycles for the suspended CNTs, and (**b**) for the non-suspended CNTs.

The humidity sensing mechanism of CNTs has been studied and discussed extensively in previous works [10]. The sensing is primarily due to charge transfer within the CNT networks. The CNTs are inherently p-type. As the water molecules donate electrons to the CNTs as a result, the number of primary charge carriers in the CNTs (holes) decreases and thus the resistance increases in the sensor. The sensitivity factor is given by [10,22]:

$$S = \frac{R_H - R_0}{R_0} \times 100\%,\qquad(1)$$

where R_H is the resistance at any given humidity, R_0 is the starting resistance at 15% RH. The sensitivity factor comparison of the suspended and non-suspended sensors is shown in Figure 5. At 98% humidity, the sensitivity factor of the suspended sensor is 172.9% compared to 64.2% for the non-suspended sensor. This outlines a 2.7-fold improvement for the suspended structure. The slope of the graph, which represents the percentage change in the resistance from the initial resistance per % RH change, for the suspended sensor is 2.05 compared to 0.68 for the non-suspended sensor, meaning the sensitivity of the suspended sensor is 3-fold greater than the non-suspended sensor. Demonstrably, the suspended sensor shows better performance in all aspects and thus is a better prospect for CNT-based humidity sensors. The acidic functionalization of CNTs shortens the CNTs, making ends open and introduces side wall defects, which facilitate water adsorption. The increased water adsorption capacity combined with shortened pathways for the charge carriers result in faster response times in the functionalized CNT sensor [23].

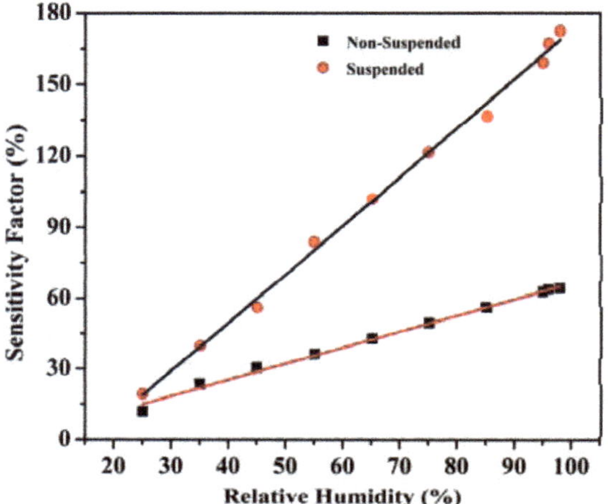

Figure 5. Sensitivity factors of suspended CNTs vs. non-suspended CNTs.

3.3. Temperature Study

The resistance of the suspended devices was measured at different temperatures to gauge the performance under different thermal operating conditions. In Figure 6, the resistance of the suspended and non-suspended sensors as a function of temperature is plotted. As can be seen, the resistance of the devices decreases with an increase in temperature. This is attributed to a decrease in the number of water molecules available to interact with the CNT networks at higher temperatures. Moreover, the thermal energy of the electrons provided by the water molecules increases with increasing temperature. The imparted thermal energy is enough for the electrons to overcome the potential barrier between the CNTs in the network, which results in a decrease in the film's resistance [24,25].

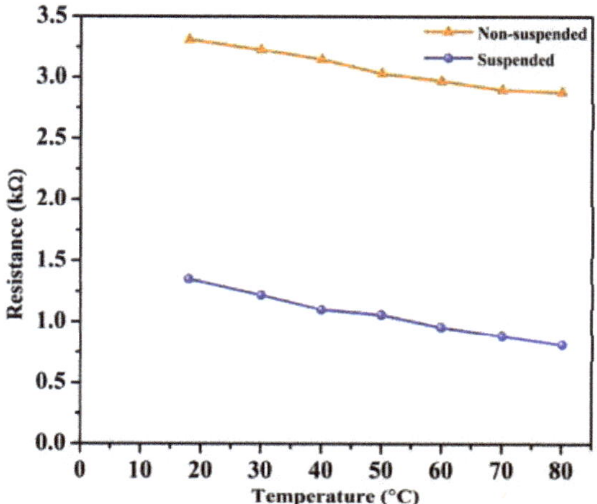

Figure 6. Resistance vs. temperature for suspended and non-suspended CNTs.

3.4. Long-Term Stability

The long-term stability of a sensor is a major factor in determining its performance. To gauge long-term stability, the resistance of the suspended device was measured at regular intervals of seven to 10 days. Figure 7a shows the resistance value of the sensor at two different humidity levels over a 45-day period. This demonstrates the stability of the sensor over extended periods as the resistance fluctuates less than 3.32% over the timespan of the test. To gauge the sensor recovery time after prolonged exposure, the sensor was soaked in a humid environment at 98% RH for 24 h.

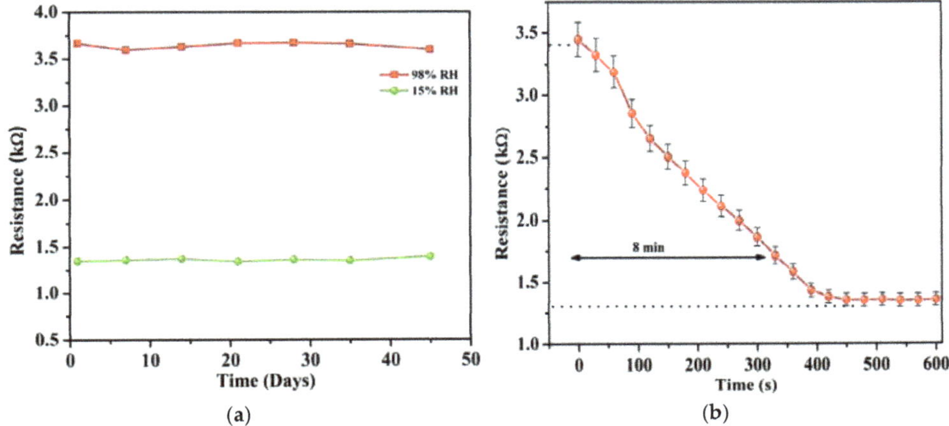

Figure 7. (a) Resistance as a function of time for suspended CNTs, indicating device stability, and (b) recovery time to 15% RH of the suspended CNTs after exposure to 98% RH for 24 h.

The humidity was then flushed out and the sensor response was recorded. Figure 7b shows the graph of the sensor recovery after prolonged exposure. The sensor took approximately 8 min to recover from 90% to 10% of its original resistance. In addition, no permanent shift in the sensor resistance was exhibited after recovery. Table 1 shows a comparison of the response and recovery times of few of the different CNT humidity sensors in literature. The presented sensor exhibits good response and recovery times compared to other works. In addition, the device also exhibits enhanced sensitivity factor when compared to the other works and a 3-fold improved sensitivity than a non-suspended sensor.

Table 1. Comparison of the response and recovery times of different CNT humidity sensors.

Publication	Sensing Material	Response Time (s)	Recovery Time (s)	Sensitivity Factor (%)
This work	Functionalized Single-walled CNTs	12	47	172.9
Jung et al. [22]	Metal oxide coated CNTs	30	25	~60
Cao et al. [10]	Functionalized Multiwalled CNTs	50	140	124
Mudimela et al. [16]	Single-walled CNTs	180	240	N/A
Zhang et al. [26]	Multiwalled CNTs	60	70	80
Moraes et al. [27]	Functionalized Multiwalled CNTs	3	90	135
Chen et al. [5]	Multiwalled CNTs	16	8	29.9
Arunachalam et al. [19]	Single-walled CNTs	290	510	246.9

4. Conclusions

In this work, a humidity sensor based on a suspended carbon nanotube has been demonstrated. The suspended sensor showed good performance with near-zero hysteresis, fast response and recovery times, and exhibited good long-term stability. To demonstrate the advantages of using a suspended architecture, the performance of suspended CNTs was compared to a non-suspended sensor under similar testing conditions. Evidently, the suspended sensors showed superior humidity sensing characteristics such as near-zero hysteresis, and significantly improved sensitivity owing to lack of substrate effects and more surface area for adsorption. Moreover, the work improved the response time of the devices as compared to our previous work in [19] by more than 10-fold due to the chemical functionalization of the nanotubes making them more sensitive to the water molecules. The suspended humidity CNT sensor presented here also compares favorably in terms of response and recovery times to other literary works. The proposed sensor design thus has the potential of yielding high-performance humidity sensors that are suitable to integration with integrated circuits.

Author Contributions: S.A. did all the experimental work, data acquisition, and analysis. F.N. and R.I. contributed expertise, direction, materials, and experimental tools.

Funding: The authors wish to thank the Natural Science and Engineering Research Council (NSERC) of Canada for its financial support of this work.

Conflicts of Interest: The authors declare no conflict of interest.

References

1. Iijima, S. Helical microtubules of graphitic carbon. *Nature* **1991**, *354*, 56. [CrossRef]
2. Cantalini, C.; Valentini, L.; Armentano, I.; Kenny, J.M.; Lozzi, L.; Santucci, S. Carbon nanotubes as new materials for gas sensing applications. *J. Eur. Ceram. Soc.* **2004**, *24*, 1405–1408. [CrossRef]
3. Wang, Y.; Yeow, J.T.W. A Review of Carbon Nanotubes-Based Gas Sensors. *J. Sens.* **2009**, *2009*, 24. [CrossRef]
4. Liang, Y.X.; Chen, Y.J.; Wang, T.H. Low-resistance gas sensors fabricated from multiwalled carbon nanotubes coated with a thin tin oxide layer. *Appl. Phys. Lett.* **2004**, *85*, 666–668. [CrossRef]
5. Chen, W.P.; Zhao, Z.G.; Liu, X.W.; Zhang, Z.X.; Suo, C.G. A Capacitive Humidity Sensor Based on Multi-Wall Carbon Nanotubes (MWCNTs). *Sensors* **2009**, *9*, 7431–7444. [CrossRef] [PubMed]
6. Snow, E.S.; Perkins, F.K.; Houser, E.J.; Badescu, S.C.; Reinecke, T.L. Chemical Detection with a Single-Walled Carbon Nanotube Capacitor. *Science* **2005**, *307*, 1942–1945. [CrossRef] [PubMed]
7. Penza, M.; Rossi, R.; Alvisi, M.; Aversa, P.; Cassano, G.; Suriano, D.; Benetti, M.; Cannata, D.; Di Pietrantonio, F.; Verona, E. SAW Gas sensors with carbon nanotubes films. In Proceedings of the 2008 IEEE Ultrasonics Symposium, Beijing, China, 2–5 November 2008.
8. Ma, Y.; Ma, S.; Wang, T.; Fang, W. Air-flow sensor and humidity sensor application to neonatal infant respiration monitoring. *Sens. Actuators A Phys.* **1995**, *49*, 47–50. [CrossRef]
9. Farahani, H.; Wagiran, R.; Hamidon, M. Humidity Sensors Principle, Mechanism, and Fabrication Technologies: A Comprehensive Review. *Sensors* **2014**, *14*, 7881–7939. [CrossRef] [PubMed]
10. Cao, C.L.; Hu, C.G.; Fang, L.; Wang, S.X.; Tian, Y.S.; Pan, C.Y. Humidity Sensor Based on Multi-Walled Carbon Nanotube Thin Films. *J. Nanomater.* **2011**, *2011*, 5. [CrossRef]
11. Tsai, J.T.; Lu, C.C.; Li, J.G. Fabrication of humidity sensors by multi-walled carbon nanotubes. *J. Exp. Nanosci.* **2010**, *5*, 302–309. [CrossRef]
12. Tang, Q.Y.; Chan, Y.C.; Zhang, K. Fast response resistive humidity sensitivity of polyimide/multiwall carbon nanotube composite films. *Sens. Actuators B Chem.* **2011**, *152*, 99–106. [CrossRef]
13. Liu, L.; Ye, X.; Wu, K.; Han, R.; Zhou, Z.; Cui, T. Humidity Sensitivity of Multi-Walled Carbon Nanotube Networks Deposited by Dielectrophoresis. *Sensors* **2009**, *9*, 1714–1721. [CrossRef] [PubMed]
14. Yeow, J.T.W.; She, J.P.M. Carbon nanotube-enhanced capillary condensation for a capacitive humidity sensor. *Nanotechnology* **2006**, *17*, 5441. [CrossRef]
15. Zhao, Z.G.; Liu, X.W.; Chen, W.P.; Li, T. Carbon nanotubes humidity sensor based on high testing frequencies. *Sens. Actuators A Phys.* **2011**, *168*, 10–13. [CrossRef]
16. Mudimela, P.R.; Grigoras, K.; Anoshkin, I.V.; Varpula, A.; Ermolov, V.; Anisimov, A.S.; Nasibulin, A.G.; Novikov, S.; Kauppinen, E.I. Single-Walled Carbon Nanotube Network Field Effect Transistor as a Humidity Sensor. *J. Sens.* **2012**, *2012*, 7. [CrossRef]
17. Chikkadi, K.; Muoth, M.; Liu, W.; Maiwald, V.; Hierold, C. Enhanced signal-to-noise ratio in pristine, suspended carbon nanotube gas sensors. *Sens. Actuators B Chem.* **2014**, *196*, 682–690. [CrossRef]

18. Muoth, M.; Helbling, T.; Durrer, L.; Lee, S.W.; Roman, C.; Hierold, C. Hysteresis-free operation of suspended carbon nanotube transistors. *Nat. Nanotechnol.* **2010**, *5*, 589. [CrossRef] [PubMed]

19. Arunachalam, S.; Gupta, A.A.; Izquierdo, R.; Nabki, F. Suspended Carbon Nanotubes for Humidity Sensing. *Sensors* **2018**, *18*, 1655. [CrossRef]

20. Cao, J.; Wang, Q.; Dai, H. Electron transport in very clean, as-grown suspended carbon nanotubes. *Nat. Mater.* **2005**, *4*, 745. [CrossRef]

21. Sapmaz, S.; Jarillo-Herrero, P.; Blanter, Y.M.; Dekker, C.; Van Der Zant, H.S.J. Tunneling in Suspended Carbon Nanotubes Assisted by Longitudinal Phonons. *Phys. Rev. Lett.* **2006**, *96*, 026801. [CrossRef]

22. Jung, D.; Kim, J.; Lee, G.S. Enhanced humidity-sensing response of metal oxide coated carbon nanotube. *Sens. Actuators A Phys.* **2015**, *223*, 11–17. [CrossRef]

23. Zhang, J.; Zou, H.; Qing, Q.; Yang, Y.; Li, Q.; Liu, Z.; Guo, X.; Du, Z. Effect of Chemical Oxidation on the Structure of Single-Walled Carbon Nanotubes. *J. Phys. Chem. B* **2003**, *107*, 3712–3718. [CrossRef]

24. Benchirouf, A.; Palaniyappan, S.; Ramalingame, R.; Raghunandan, P.; Jagemann, T.; Müller, C.; Hietschold, M.; Kanoun, O. Electrical properties of multi-walled carbon nanotubes/PEDOT:PSS nanocomposites thin films under temperature and humidity effects. *Sens. Actuators B Chem.* **2016**, *224*, 344–350. [CrossRef]

25. Zhang, H.L.; Li, J.F.; Zhang, B.P.; Yao, K.F.; Liu, W.S.; Wang, H. Electrical and thermal properties of carbon nanotube bulk materials: Experimental studies for the 328-958 K temperature range. *Phys. Rev. B* **2007**, *75*, 205407. [CrossRef]

26. Zhang, Y.; Yu, K.; Xu, R.; Jiang, D.; Luo, L.; Zhu, Z. Quartz crystal microbalance coated with carbon nanotube films used as humidity sensor. *Sens. Actuators A Phys.* **2005**, *120*, 142–146. [CrossRef]

27. Moraes, M.C.; Galeazzo, E.; Peres, H.E.; Ramirez-Fernandez, F.J.; Dantas, M.O. Development of Fast Response Humidity Sensors Based on Carbon Nanotubes. Available online: https://www.researchgate.net/profile/Michel_Dantas/publication/268742675_Development_of_Fast_Response_Humidity_Sensors_Based_on_Carbon_Nanotubes/links/54c7e8900cf289f0cece471d.pdf (accessed on 5 February 2019).

Article

Design and Mechanical Sensitivity Analysis of a MEMS Tuning Fork Gyroscope with an Anchored Leverage Mechanism

Zezhang Li [1], Shiqiao Gao [1,*], Lei Jin [2], Haipeng Liu [1], Yanwei Guan [3] and Shigang Peng [1]

1   State Key Laboratory of Explosion Science and Technology, Beijing Institute of Technology, Beijing 100081, China
2   School of Mechatronical Engineering, Beijing Institute of Technology, Beijing 100081, China
3   Beijing Institute of Control and Electronic Technology, Beijing 100038, China
*   Correspondence: gaoshq@bit.edu.cn; Tel.: +86-10-6891-1631

Received: 28 May 2019; Accepted: 6 August 2019; Published: 7 August 2019

Abstract: This paper presents the design and analysis of a new micro-electro-mechanical system (MEMS) tuning fork gyroscope (TFG), which can effectively improve the mechanical sensitivity of the gyroscope sense-mode by the designed leverage mechanism. A micromachined TFG with an anchored leverage mechanism is designed. The dynamics and mechanical sensitivity of the design are theoretically analyzed. The improvement rate of mechanical sensitivity (IRMS) is introduced to represent the optimization effect of the new structure compared with the conventional one. The analytical solutions illustrate that the IRMS monotonically increases with increased stiffness ratio of the power arm (SRPA) but decreases with increased stiffness ratio of the resistance arm (SRRA). Therefore, three types of gyro structures with different stiffness ratios are designed. The mechanical sensitivities increased by 79.10%, 81.33% and 68.06% by theoretical calculation. Additionally, FEM simulation demonstrates that the mechanical sensitivity of the design is in accord with theoretical results. The linearity of design is analyzed, too. Consequently, the proposed new anchored leverage mechanism TFG offers a higher displacement output of sense mode to improve the mechanical sensitivity.

Keywords: mechanical sensitivity; tuning fork gyroscope; anchored leverage mechanism; stiffness ratio; coordinate transformation method

---

1. Introduction

A micro-electro-mechanical system (MEMS) gyroscope is a kind of inertial sensor used to detect the attitude angle and angular rate. It is based on an energy conversion of two vibrational modes due to the Coriolis effect [1,2]. With the rapid development of MEMS technology, the MEMS gyroscope has long been considered as an attractive and dynamic inertial sensor for a wide variety of applications, including the automotive industry, robotics, consumer electronics and high-volume military uses (0.1 to 100°/h). Compared with conventional gyroscopes, MEMS gyroscopes have many advantages, such as tiny volume, light weight, low power consumption, low cost, and the possibility of batch fabrication [3,4].

A common type of MEMS vibratory gyroscope is implemented as a tuning fork gyroscope (TFG). It is composed of two identical tines and two coupling mechanisms for synchronization of the anti-phase drive mode and anti-phase sense mode. The advantage of the TFG is that it can cancel the external common mode effect by applying a differential Coriolis detection between the two tines [5–9]. The sensitivity of the TFG is the biggest challenge in terms of superior performance.

The sensitivity of a micromachined TFG depends on two factors, peripheral circuit gain and the gyro's own mechanical sensitivity. Much effort has already been made to improve the circuit gain and precision, and many achievements have been obtained [10,11]. Improving the sensitivity of the system only by enlarging the amplification of the peripheral circuit is meaningless. Since the Coriolis signal and the noise are often amplified at the same time, the system could not improve the signal-to-noise ratio (SNR). Therefore, to improve the sensitivity of vibratory gyroscopes, it is essential to achieve high mechanical sensitivity. The vibrational amplitude greatly improves when the work frequency equals the resonant frequency. Therefore, the displacement of differential tines is maximized and the mechanical sensitivity can be effectively increased when the two work modes of the gyroscope have the same resonant frequency (i.e., they are mode matched) [12,13]. It is difficult to completely match the resonant frequencies of the two modes through structural design due to fabrication imperfections. Therefore, mode matching through electrostatic tuning is often adopted [14–18]. Minimizing substrate energy dissipation and vacuum packaging can improve sensitivity owing to a high quality factor [19–22]. On the other hand, many studies on vibration type angular rate sensors in materials group them into bulk Si and polycrystalline Si [23]; some researchers focus on quartz gyroscopes based on material characteristics [24].

Amplification mechanisms are gaining importance in MEMS devices where motion reliability, precision, and sensitivity are needed. Researchers all over the world have proposed different types of amplification mechanisms, such as bridge-type, lever-type and four-bar linkage mechanisms [25–30]. The amplification mechanism has been applied to MEMS resonant output gyroscope (ROG) and MEMS accelerometer due to the force or displacement amplification effects [31,32]. The focus of this paper is on how to improve the mechanical sensitivity of a MEMS gyroscope by using a leverage mechanism.

In this paper, a new kind of micromachined TFG with an anchored leverage mechanism is proposed to analyze mechanical sensitivity. A conventional one with direct connection between sense-mode frame and proof mass is introduced at the same time for comparison. Detailed descriptions of the two design structures are provided in Section 2. Section 3 establishes and solves the dynamic equations for the response of the two architectures. FEM simulation of the stiffness of different springs on sense-mode and comparisons between simulation and analytical solutions are discussed and the linearity of the design is analyzed in Section 4. In Section 5, the discussion is given. Section 6 concludes the paper with a summary.

2. Architecture Design

In this paper, a novel MEMS tuning fork gyroscope with an anchored leverage mechanism is designed. The architecture of type A, as discussed in detail in Figure 1a, is a dual-mass structure that consists of an anchored diamond coupled spring, two sense levers, and two identical tines. As a conventional structure, type B is the same as type A, except for the connection between the proof mass and the sense-mode frame, as depicted in Figure 1b.

Both type A and type B are completely symmetrical double structured–decoupled architectures. Each tine contains a proof mass, two drive-mode frames and two sense-mode frames supported by symmetrical springs, and type A has an additional lever mechanism. In order to improve the robustness of the mode match between drive mode and sense mode, the springs are located on the same axis and can resist any temperature change of the resonance frequency, except the supporting lever springs and the anchored diamond coupled spring. All springs in the design are U-shaped in series or in parallel, which can reduce their axial stress to a greater extent, so that there is a linear relationship between force and displacement, which matches the electrodes by the variable-area capacitance mechanism.

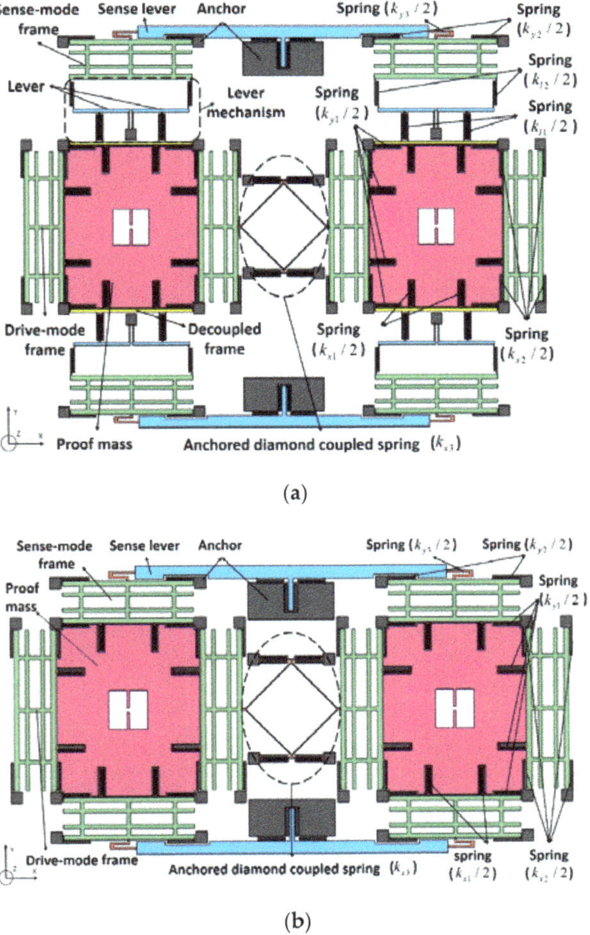

(**a**)

(**b**)

Figure 1. Schematic of designed tuning fork gyroscopes (TFGs): (**a**) type A, (**b**) type B.

3. Theoretical Analysis

3.1. Kinematic Analysis of Type A and B

The spring in Figure 1 represents the stiffness of the drive beams, the sense beams and the decoupling beams in the design of this paper. The stiffness of these beams is much less than that of the lever, frame and proof mass. On the other hand, the mass of the lever, frame, and proof mass is much larger than that of the beam. Therefore, the lever, frame and proof mass are simplified to inelastic mass and beam to massless spring. It is worth mention here that the lumped-parameter model used in this study is based on linear elastic material behavior, which is very widely used in the research of TFG [6,14,21,33–35]. Furthermore, the nonlinear analysis of the overall structure will be discussed in detail later. However, when the stiffness and mass of the beam are relatively large compared to other structures, some errors will occur and will not be adapted to this assumption.

According to Figure 1a, we can obtain the four-degrees-of-freedom (4-dof) coupling vibration model of the MEMS tuning fork gyroscope with an anchored leverage mechanism, as illustrated in Figure 2.

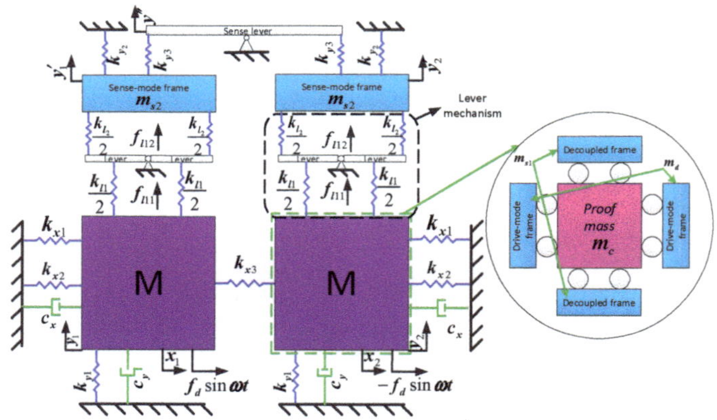

Figure 2. Model of type A.

In operation, two proof masses and their respective drive mechanisms are electrostatically driven into anti-phase motion with the same amplitude along the drive direction by driving voltages imposed across the differential lateral comb electrodes on the drive mechanism. When an angular rate Ω_z is applied, the anti-phase Coriolis acceleration of the proof mass induces linear anti-phase motions that are capacitively detected using differential parallel plate electrodes on the sense mechanism along the y-axis. The input angular rate Ω_z can be calculated by the differential output.

Ideally, the structure is entirely symmetrical. Then, the two proof masses systems and damping coefficients are equal. The dynamics in the direction of sense mode are governed by the following:

Proof mass:

$$\begin{cases} (m_{s1} + m_c)\ddot{y}_1 + c_y\dot{y}_1 + \left(k_{y1} + k_{l1}\right)y_1 + f_{l11} = -2m_c\Omega_z\dot{x}_1 \\ (m_{s1} + m_c)\ddot{y}_2 + c_y\dot{y}_2 + \left(k_{y1} + k_{l1}\right)y_2 + f_{l21} = -2m_c\Omega_z\dot{x}_2 \end{cases} \tag{1}$$

where m_c, m_{s1} and m_{s2} are the mass of the proof mass, decoupled frame and sense-mode frame, respectively. c_y is the damping coefficient of each tine in sense direction. k_{y1} is the stiffness of the spring connected to m_c and the drive-mode frame. k_{l1} represents the stiffness of the spring connected to decoupled frame and lever in the sense direction. f_{l11} and f_{l11} represent the force exerted on the lever by the decoupled frame in the left and right tines. Ω_z is the angular rate and the drive velocity $\dot{x}_1$ can be defined as:

$$\begin{cases} \dot{x}_1(t) = -\dfrac{Q_{x\_an}f_d\omega_{x\_an}}{k_{x\_an}}\sin\omega_{x\_an}t \\ \dot{x}_2(t) = \dfrac{Q_{x\_an}f_d\omega_{x\_an}}{k_{x\_an}}\sin\omega_{x\_an}t \end{cases} \tag{2}$$

Sense-mode frame:

$$\begin{cases} m_{s2}\ddot{y}'_1 + \left(k_{y2} + k_{l2}\right)y'_1 + k_{y3}\left(y'_1 - y''\right) = f_{l12} \\ m_{s2}\ddot{y}'_2 + \left(k_{y2} + k_{l2}\right)y'_1 + k_{y3}\left(y'_2 + y''\right) = f_{l22} \end{cases} \tag{3}$$

where k_{y1} is the stiffness of the spring connected to the anchor and the sense-mode frame; k_{l1} represents the stiffness of the spring connected to the sense-mode frame and lever in the sense direction; k_{y3} is the stiffness in the direction of the spring connected to the sense lever and sense-mode frame; and f_{l11} and f_{l11} represent the force exerted on the sense-mode frame by the lever in the left and right tines, respectively.

The leverage mechanism and coordinate relationships are as follows:

$$\begin{cases} f_{l11} = Bf_{l12} \\ f_{l21} = Bf_{l22} \\ y' = By \end{cases} \tag{4}$$

where B is the leverage rate (LR), $B > 1$.

Sensing coupling frame:

$$k_{y3}(y'_1 - y'') = k_{y3}(y'_2 + y'') \tag{5}$$

According to the above equations, the kinematic analysis of the dual-mass gyroscope in the sense direction can be expressed as:

$$(m_{s1} + m_c + B^2 m_{s2})\ddot{y}_1 + c_y\dot{y}_1 + \left(k_{y1} + k_{l1} + B^2 k_{y2} + B^2 k_{l2} + \frac{B^2 k_{y3}}{2}\right)y_1 + \frac{B^2 k_{y3}}{2}y_2 = -2m_c\Omega_z\dot{x}_1 \tag{6}$$

$$(m_{s1} + m_c + B^2 m_{s2})\ddot{y}_2 + c_y\dot{y}_2 + \left(k_{y1} + k_{l1} + B^2 k_{y2} + B^2 k_{l2} + \frac{B^2 k_{y3}}{2}\right)y_2 + \frac{B^2 k_{y3}}{2}y_1 = -2m_c\Omega_z\dot{x}_2 \tag{7}$$

Subtracting Equation (7) from Equation (6), we obtain

$$(m_{s1} + m_c + B^2 m_{s2})(\ddot{y}_1 - \ddot{y}_2) + c_y(\dot{y}_1 - \dot{y}_2) + \left(k_{y1} + k_{l1} + B^2 k_{y2} + B^2 k_{l2}\right)(y_1 - y_2) = 2f_c \sin\omega_{x\_an}t \tag{8}$$

where f_c is the Coriolis force at angulate rate Ω_z, which can be given as:

$$f_c = \frac{2m_c\Omega_z Q_{x\_an} f_d \omega_{x\_an}}{k_{x\_an}} \tag{9}$$

where f_d is the driving force, and ω_{x_an}, k_x, and Q_x are the resonant frequency, total stiffness, and quality factor, respectively, in the anti-mode.

Adding Equations (6) and (7), we obtain:

$$(m_{s1} + m_c + B^2 m_{s2})(\ddot{y}_1 + \ddot{y}_2) + c_y(\dot{y}_1 + \dot{y}_2) + \left(k_{y1} + k_{l1} + B^2 k_{y2} + B^2 k_{l2} + B^2 k_{y3}\right)(y_1 + y_2) = 0 \tag{10}$$

Since the vibration output of in- and anti-phase modes needs to be explored, a coordinate transformation is made as follows:

$$y_{an} = y_1 - y_2, \ y_{in} = y_1 + y_2 \tag{11}$$

Substituting Equation (11) into Equations (8) and (9), we obtain:

$$\begin{cases} \ddot{y}_{an} + \dfrac{\omega_{y\_an}}{Q_{y\_an}}\dot{y}_{an} + \omega^2_{y\_an}y_{an} = 2f_c \sin\omega_{x\_an}t \\ \ddot{y}_{in} + \dfrac{\omega_{y\_in}}{Q_{y\_in}}\dot{y}_{in} + \omega^2_{y\_in}y_{in} = 0 \end{cases} \tag{12}$$

where $\omega_{y\_an} = \sqrt{\dfrac{k_{y1}+k_{l1}+B^2k_{y2}+B^2k_{l2}}{m_y}}$, $\omega_{y\_in} = \sqrt{\dfrac{k_{y1}+k_{l1}+B^2k_{y2}+B^2k_{l2}+B^2k_{y3}}{m_y}}$, $Q_{y\_an} = \dfrac{m_y\omega_{y\_an}}{c_y}$, $Q_{y\_in} = \dfrac{m_y\omega_{y\_in}}{c_y}$, and $m_y = m_{s1} + m_c + B^2 m_{s2}$, in which ω_{y_an} and ω_{y_in} are the defined resonant frequencies and Q_{y_an} and Q_{y_in} are the quality factors of the anti- and in-phase motions, respectively, and m_y is the total mass in the sense direction.

Equation (12) can be represented as a matrix:

$$M\ddot{y} + C\dot{y} + Ky = F_c \sin\omega t \tag{13}$$

where $M = \begin{bmatrix} 1 & \\ & 1 \end{bmatrix}$, $C = \begin{bmatrix} \frac{\omega_{y\_an}}{Q_{y\_an}} & \\ & \frac{\omega_{y\_in}}{Q_{y\_in}} \end{bmatrix}$, $K = \begin{bmatrix} \omega^2_{y\_an} & \\ & \omega^2_{y\_in} \end{bmatrix}$, $F_c = \begin{bmatrix} \frac{2f_c}{m_y} \\ 0 \end{bmatrix}$, and $y = \begin{bmatrix} y_{an} \\ y_{in} \end{bmatrix}$.

The natural frequency can be obtained by using the characteristic equation:

$$\omega_1^2 = \omega_{y\_an}^2 = \frac{k_{y1}+k_{l1}+B^2k_{y2}+B^2k_{l2}}{m_y}$$
$$\omega_2^2 = \omega_{y\_in}^2 = \frac{k_{y1}+k_{l1}+B^2k_{y2}+B^2k_{l2}+B^2k_{y3}}{m_y} \tag{14}$$

where ω_1 and ω_2 are the first- and second-order resonant frequency, respectively.

The modal superposition technique is used to acquire the steady-state response by solving Equation (13):

$$y(t) = \frac{f_c\beta_1}{m_y\omega_1^2}\begin{bmatrix} 1 \\ 0 \end{bmatrix}\sin(\omega_{x\_an}t - \psi_1) + \frac{f_c\beta_2}{m_y\omega_2^2}\begin{bmatrix} 0 \\ 1 \end{bmatrix}\sin(\omega_{x\_an}t - \psi_2) \tag{15}$$

where $\beta_i = \dfrac{1}{\sqrt{\left(1-\lambda_i^2\right)^2+\left(\frac{\lambda}{Q_i}\right)^2}}$, $\psi_i = \arctan\dfrac{\lambda}{(1-\lambda^2)Q_i}$, and $\lambda_i = \dfrac{\omega_{x\_an}}{\omega_i}$ are the magnification factors of amplitude, phase angle, and frequency ratio, respectively.

When $\omega = \omega_{x\_an} = \omega_{y\_an}$, we can obtain that:

$$\begin{cases} y'_1(t) = By_{an}(t) = \dfrac{2m_c\Omega_zQ_{x\_an}f_d\omega BQ_{y\_an}}{k_{y\_an}k_{x\_an}}\sin\left(\omega t - \frac{\pi}{2}\right) \\[3mm] y'_2(t) = By_{an}(t) = -\dfrac{2m_c\Omega_zQ_{x\_an}f_d\omega BQ_{y\_an}}{k_{y\_an}k_{x\_an}}\sin\left(\omega t - \frac{\pi}{2}\right) \end{cases} \tag{16}$$

According to Equation (16), the differential detection output of the tuning fork micromechanical gyroscope is:

$$y_{dA} = y'_1(t) - y'_2(t) = \frac{2f_cBQ_{y\_an}}{k_{y\_an}}\sin\left(\omega t - \frac{\pi}{2}\right) = \frac{4m_c\Omega_zQ_{x\_an}f_d\omega BQ_{y\_an}}{k_{y\_an}k_{x\_an}}\sin\left(\omega t - \frac{\pi}{2}\right) \tag{17}$$

The mechanical sensitivity of the type A architecture can be obtained from Equation (17):

$$S_{ma} = \frac{y'_1(t) - y'_2(t)}{\Omega_z} = \frac{4m_cQ_{x\_an}f_d\omega BQ_{y\_an}}{k_{y\_an}k_{x\_an}}\sin\left(\omega t - \frac{\pi}{2}\right) \tag{18}$$

The 4-dof coupling vibration model of the MEMS gyroscope with equal displacement capacitance detection is shown in Figure 3.

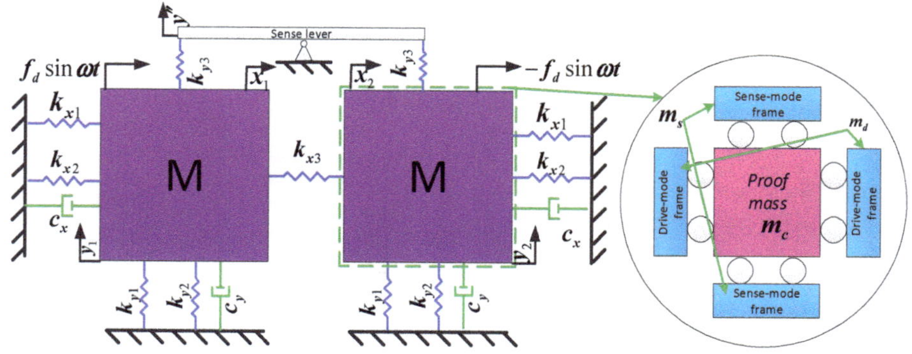

Figure 3. Model of type B.

The differential detection output and mechanical sensitivity of the type B architecture can be acquired by the same technique:

$$y_{dB} = y_{1b}(t) - y_{2B}(t) = \frac{2f_c Q_{yb\_an}}{k_{yb\_an}} \sin\left(\omega t - \frac{\pi}{2}\right) = \frac{4m_c \Omega_z Q_{x\_an} f_d \omega B Q_{yb\_an}}{k_{yb\_an} k_{x\_an}} \sin\left(\omega t - \frac{\pi}{2}\right) \quad (19)$$

$$S_{mb} = \frac{4m_c Q_{x\_an} f_d \omega Q_{yb\_an}}{k_{yb\_an} k_{x\_an}} \sin\left(\omega t - \frac{\pi}{2}\right) \quad (20)$$

where $Q_{yb\_an} = \frac{m_{yb} \omega_{yb\_an}}{c_{yb}}$, $k_{yb\_an} = k_{y1} + k_{y2}$, $m_{yb} = m_{s2} + m_c$, and $\omega_{yb\_an} = \sqrt{\frac{k_{y1} + k_{y2}}{m_{s2} + m_c}}$ are the quality factor, total stiffness, total mass, and resonant frequency of the anti-phase mode in the drive direction, respectively.

3.2. Optimization Analysis of LR

In order to obtain a more substantial improvement of mechanical sensitivity, it is effective to reduce k_{y_an} and increase B as much as possible.

Here, the dimensionless parameters α, K, and β are defined in terms of stiffness ratio (SR): α is the stiffness ratio of the power arm (SRPA) and β is the stiffness ratio of the resistance arm (SRRA). The three parameters are given by:

$$\alpha = \frac{k_{l1}}{k_{y1}}, \quad K = \frac{k_{y2}}{k_{y1}} \text{ and } \beta = \frac{k_{l2}}{k_{y1}} \quad (21)$$

Obviously, tiny LR will lead to an insignificant amplification effect of leverage, which will not be important in truly amplifying the mechanical sensitivity. On the other hand, extremely large LR will increase the detection mode stiffness, thus making it difficult for the tiny Coriolis force to drive the detection mode effectively. Without considering the effect of the quality factor during the design phase, according to Equation (18), we obtain:

$$\frac{\partial S_{ma}}{\partial B} = \frac{(1 + \alpha) - B^2 (K + \beta)}{(1 + \alpha + B^2 (K + \beta))^2} \cdot \frac{4m_c Q_{x\_an} f_d \omega Q_{y\_an}}{k_{y1} k_{x\_an}} \sin\left(\omega t - \frac{\pi}{2}\right) \quad (22)$$

When $\frac{\partial S_{ma}}{\partial B} = 0$, we can obtain that:

$$B = \sqrt{\frac{1 + \alpha}{K + \beta}} \quad (23)$$

Equation (23) shows that an optimal solution to this problem exists and different parameters have different effects on the optimal leveraged magnification. Furthermore, the square of B is directly proportional to α and inversely proportional to K and β.

3.3. Analysis of IRMS

Type A and type B adopt the same structure and springs, except for the connection between the proof mass and detection frame, and the same fabrication and packaging technique. It is assumed that the damping ratios of the two types of TFG are equal.

The dimensionless parameters η and M are defined by:

$$\eta = \frac{S_{ma} - S_{mb}}{S_{mb}} = \frac{y_{dA} - y_{dB}}{y_{dB}}, \quad M = \frac{m_{s2}}{m_c} \quad (24)$$

where η denotes the improvement rate of mechanical sensitivity (IRMS) and M denotes the mass ratio (MR).

Substituting Equations (18) and (20) into Equation (24), we obtain:

$$\eta = \sqrt{\frac{((1+\alpha)M + K + \beta)(1+K)}{2(1+M)(K+\beta)^2}} - 1 \tag{25}$$

It can be seen that the IRMS is mainly determined by the stiffness ratios K, α, and β and mass ratio M. By solving the partial derivative of function η with respect to these variables, we obtain that:

$$\frac{\partial \eta}{\partial K} = -\frac{1}{4} \cdot \frac{(1+\alpha)(K-\beta+2)M - \sqrt{2}(\beta-1)(K+\beta)}{(K+\beta)^2 \sqrt{((1+\alpha)M + K + \beta)(1+M)(1+K)}} \tag{26}$$

$$\frac{\partial \eta}{\partial \alpha} = \frac{\sqrt{2}}{4} \cdot \frac{M\sqrt{1+K}}{(K+\beta)\sqrt{(1+M)(M(1+\alpha) + K + \beta)}} \tag{27}$$

$$\frac{\partial \eta}{\partial \beta} = -\frac{\sqrt{2}}{4} \cdot \frac{(2M(1+\alpha) + K + \beta)\sqrt{1+K}}{(K+\beta)^2 \sqrt{(1+M)(M(1+\alpha) + K + \beta)}} \tag{28}$$

$$\frac{\partial \eta}{\partial M} = -\frac{\sqrt{2}}{4} \cdot \frac{(K+\beta-1+\alpha)\sqrt{1+K}}{(K+\beta)\sqrt{(1+M)^3(M(1+\alpha) + K + \beta)}} \tag{29}$$

In general, the values of the stiffness ratios K, α, and β and mass ratio M are greater than 0 and less than 1. From Equations (26)–(29), we can find that:

$$\frac{\partial \eta}{\partial K} < 0, \ \frac{\partial \eta}{\partial \alpha} > 0, \ \frac{\partial \eta}{\partial \beta} < 0, \ \frac{\partial \eta}{\partial M} > 0 \tag{30}$$

From Equation (30), the improvement rate of mechanical sensitivity η monotonically increases with increasing α and M, but decreases with increasing K and β. Within a reasonable range, K and β are as small as possible and α and M are as large as possible. The leverage mechanism can improve the displacement of sense frame effectively and obtain a huge improvement in terms of mechanism sensitivity.

The change of M and K will lead to a corresponding change of the mechanism architecture with no leverage. Furthermore, M and K are system parameters, not just leverage mechanism parameters, which will be analyzed in later studies. Here, four architectures are designed to explore the impact of α and β on IRMS. One is based on type B architecture and the other three are based on type A architecture, defined as types A1, A2 and A3. The k_{l1} of type A2 and the k_{l2} of type A3 are slightly larger than those of type A1. This can be achieved by intentionally increasing the spring width of the leverage mechanism in the sense direction.

4. FEM Simulation and Analysis

4.1. Analysis of LR and IRMS

From Equation (21), the stiffness ratio is dependent on the stiffness of various springs in the sense direction. Therefore, simulations are carried out on the stiffness of these springs by applying a 1 μN force in the linked structure, shown in Figure 4. The length, width and height of the designed spring are 500 μm, 10 μm and 80 μm, respectively. The stiffness of the linear springs connecting the anchor to the proof mass and the sense frame is shown in Figure 4a,b, respectively. The stiffness of the beam associated with the leverage mechanism is shown in Figure 4c,d. Using the formula $k = F/x$, the stiffness k can be obtained:

$$k_{y1} = \frac{1\ \mu N}{0.0012029\ \mu m} = 831.32\ N/m,\ k_{y2} = 2 \cdot \frac{1\ \mu N}{0.0103225\ \mu m} = 193.75\ N/m$$

$$k_{l1} = 2 \cdot \frac{1\ \mu N}{0.025038\ \mu m} = 79.88\ N/m,\ k_{l2} = 2 \cdot \frac{1\ \mu N}{0.039746\ \mu m} = 50.32\ N/m \tag{31}$$

$$k_{l1\_2} = 2 \cdot \frac{1\ \mu N}{0.015593\ \mu m} = 128.26\ N/m,\ k_{l2\_3} = 2 \cdot \frac{1\ \mu N}{0.027567\ \mu m} = 72.55\ N/m$$

Substituting Equation (31) into Equations (21) and (24), we obtain:

$$K = \frac{k_{y2}}{k_{y1}} = \frac{193.75}{831.32} = 0.233063,\ \alpha = \frac{k_{l1}}{k_{y1}} = \frac{79.88}{831.32} = 0.096088,$$

$$\alpha_2 = \frac{k_{l1\_2}}{k_{y1}} = \frac{128.26}{831.32} = 0.15429 \tag{32}$$

$$\beta = \frac{k_{l2}}{k_{y1}} = \frac{50.32}{831.32} = 0.06053,\ \beta_3 = \frac{k_{l2}}{k_{y1}} = \frac{72.55}{831.32} = 0.08727$$

Substituting Equation (32) into Equation (23), we obtain:

$$B_1 = \sqrt{\frac{1+\alpha}{K+\beta}} = \sqrt{\frac{1+0.096088}{0.233063 + 0.06053}} = 1.93,\ B_2 = \sqrt{\frac{1+\alpha_2}{K+\beta}} = \sqrt{\frac{1+0.15429}{0.233063 + 0.06053}} = 1.98$$

$$B_3 = \sqrt{\frac{1+\alpha}{K+\beta_3}} = \sqrt{\frac{1+0.096088}{0.233063 + 0.08727}} = 1.85 \tag{33}$$

The mass ratio M is determined by structural design parameters; the mass of the proof mass and sense frame are $1.16e-6$ kg and $2.78e-7$ kg, so we obtain:

$$M = \frac{m_{s2}}{m_c} = \frac{2.774e-7\ kg}{1.1598e-6\ kg} = 0.23918 \tag{34}$$

Therefore, the IRMS of the MEMS tuning fork gyroscope with an anchored leverage mechanism can be calculated from Equations (25), (32) and (34):

$$\eta_1 = \sqrt{\frac{((1+\alpha)M + K + \beta)(1+K)}{2(1+M)(K+\beta)^2}} - 1 = 79.10\%,$$

$$\eta_2 = \sqrt{\frac{((1+\alpha_2)M + K + \beta)(1+K)}{2(1+M)(K+\beta)^2}} - 1 = 81.33\% \tag{35}$$

$$\eta_3 = \sqrt{\frac{((1+\alpha)M + K + \beta_3)(1+K)}{2(1+M)(K+\beta_3)^2}} - 1 = 68.06\%$$

From the above numerical analysis, it is concluded that the IRMS of type A1 is lower than that of type A2 and slightly higher than that of type A3, which is in accord with the theoretical results.

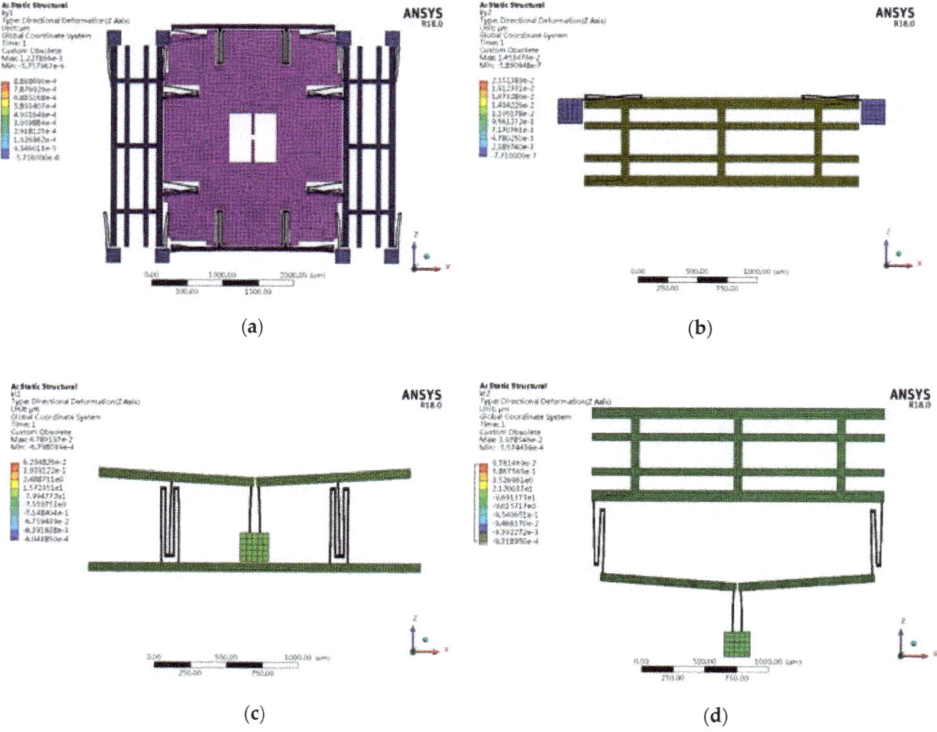

Figure 4. Stiffness of sense mode of micro-electro-mechanical (MEMS) gyroscope: (**a**) deformation of k_{y1}; (**b**) deformation of k_{y2}; (**c**) deformation of k_{l1}; (**d**) deformation of k_{l2}.

4.2. FEM Analysis of Mechanism Sensitivity

4.2.1. Modal Analysis

A structural model simulation is carried out using the ANSYS software. The mesh element is SOLID186, which is a high-order three-dimensional 20-node solid structure unit. All the structures use hexahedral meshes: the linear beam uses a small mesh, and the drive and sense frames use slightly larger meshes. All anchor and proof mass elements use larger hexahedral meshes. Types A1, A2, A3, and B have a total of 543,516, 547,996, 547,039, and 399,356 meshes, respectively. The main parameters of the structures and the material properties of silicon are shown in Tables 1 and 2, respectively.

Figure 5 shows the sense modes of types A1 and B. The anti-sense and anti-drive modes of all types are the first two order modes. The natural frequencies of the first three modes and the corresponding modes of vibration of all types are listed in Table 3.

Table 1. Main parameters of the structures.

Parameter	Value
Proof mass (μm^3)	$2300 \times 2700 \times 80$
Length of drive spring (μm)	440
Width of drive spring (μm)	10
Length of sense spring (μm)	440
Width of sense spring (μm)	10
Length of drive coupling leverage (μm)	1131
Width of drive coupling leverage (μm)	15
Length of sense coupling leverage (μm)	5450
Width of sense coupling leverage (μm)	220
Length of leverage (μm)	1050
Width of leverage (μm)	60
Lever arm length ratio	1.93
Equivalent mass of type A1 (kg)	2.2469×10^{-6}
Equivalent mass of type A2 (kg)	2.3019×10^{-6}
Equivalent mass of type A3 (kg)	2.1604×10^{-6}
Equivalent mass of type B (kg)	1.4372×10^{-6}

Table 2. Material parameters for FEM simulation.

Parameters	Young's Modulus (Pa)	Poisson's Ratio	Density (kg/m^2)
Values	1.7×10^{11}	0.28	2330

Table 3. Natural frequencies of first three modes and corresponding modes of vibration of all types.

Type	Order	1	2	3
Type A1	Frequency (Hz)	4227.3	4234.8	7914.9
	Mode of vibration	Anti-phase of drive	Anti-phase of sense	In-phase of sense
Type A2	Frequency (Hz)	4227.4	4288.9	8068.6
	Mode of vibration	Anti-phase of drive	Anti-phase of sense	In-phase of sense
Type A3	Frequency (Hz)	4227.3	4319.2	7938.4
	Mode of vibration	Anti-phase of drive	Anti-phase of sense	In-phase of sense
Type B	Frequency (Hz)	4185.2	4248	6478.9
	Mode of vibration	Anti-phase of sense	Anti-phase of drive	In-phase of sense

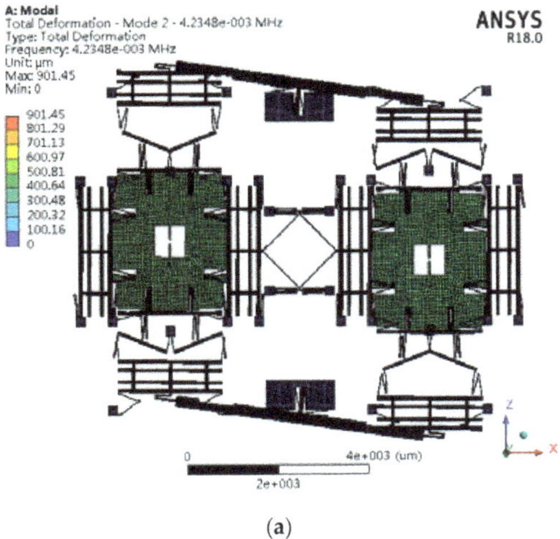

(**a**)

Figure 5. *Cont.*

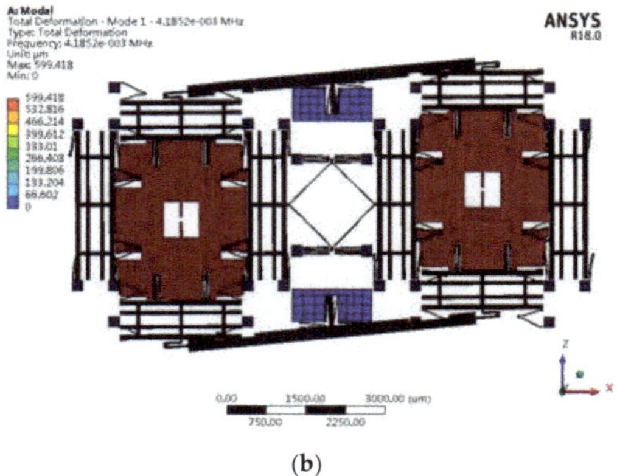

(b)

Figure 5. Anti-phase mode frequency in the sense direction of (**a**) type A1 and (**b**) type B.

It can be seen from Table 3 that the anti-sense and anti-drive modes of all types are the first two order modes. The mode order of type A with an anchored leverage mechanism does not change compared to type B. The leverage mechanism can improve the frequency of in-phase sense mode, which can decrease the influence of the common vibration on the mechanical sensitivity.

4.2.2. Sensitivity Analysis

To analyze the vibration output response, a mode-based steady-state linear dynamic analysis is performed on the design by ANSYS software to predict the linear response of a structure subjected to continuous harmonic excitation. In this simulation, the parameters of the structure, material properties of silicon, mesh division and element type are the same as those of the modal analysis. According to previous experimental results, the quality factor of MEMS gyroscopes is about 3000–8000 [33]. In this analysis, when the quality factor of type B is 4000, the quality factor of types A1, A2 and A3 can be calculated by the previous assumptions and defined as 6328, 6565 and 6605, respectively. In order to guarantee the accuracy and efficiency of the simulation, the frequency ranges of types A1 and B designs sweep from 4100 Hz to 4300 Hz based on the results of modal analysis in the previous section. The frequency step is 1 Hz. Since the amplitude response around the resonant frequency varies greatly, a harmonic response analysis with higher accuracy needs to be carried out. The frequency range is determined by the previous results. The frequency ranges of types A1 and B designs sweep from 4233.5 Hz to 4236 Hz and from 4184 Hz to 4186.5 Hz, respectively. As a result, the frequency step is 0.025 Hz. In the left and right proof mass of the design, in-phase and anti-phase simple harmonic force with a relative amplitude of 1 μN is applied. The resonance frequency of sense-mode in type A1 and type B is 4234.8 Hz and 4185.2 Hz, respectively, and the vibration amplitude of sense-mode frame is about 6.483 μm and 3.766 μm. Figure 6 displays the deformation results at 20 times magnification. The left and right sense mode frame's amplitude–frequency characteristics of the vibration of the tuning fork structure after loading are shown in Figure 7.

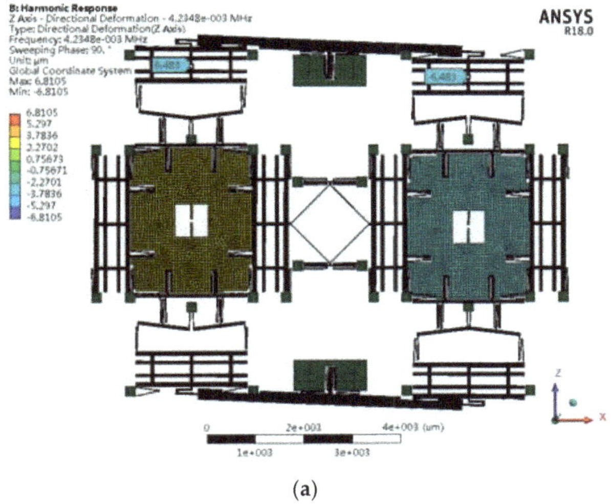

(a)

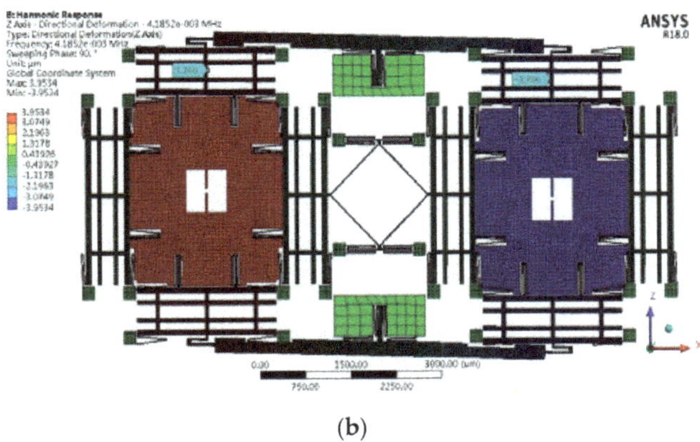

(b)

Figure 6. Directional deformation of harmonic response analysis in the sense-mode resonance frequency of (**a**) type A1 and (**b**) type B.

As depicted in Figure 7, two tines of types A and B have completely synchronized movement to eliminate the vibration output. This verifies the function as a tuning fork. The differential displacement of two tines can be obtained by computing the simulation data with MATLAB software, as shown in Figure 8.

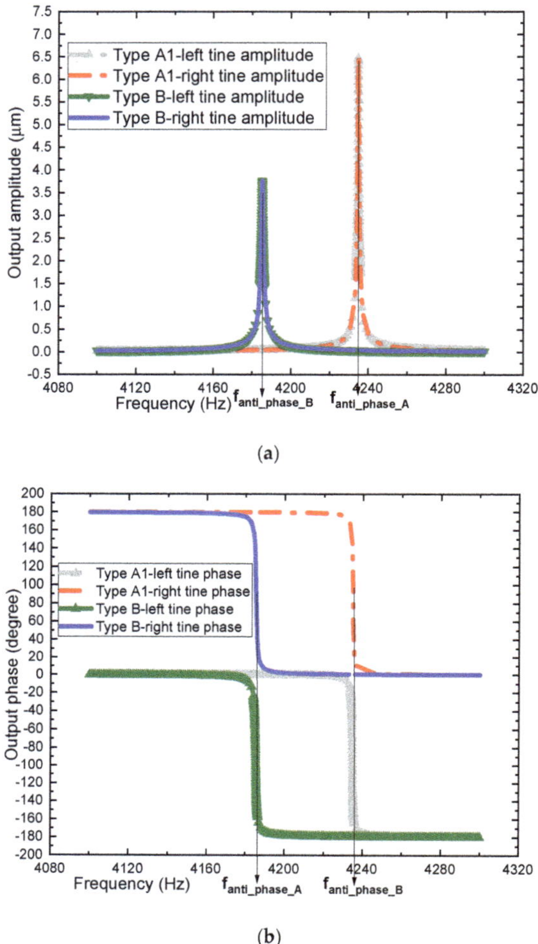

(**a**)

(**b**)

Figure 7. (**a**) Amplitude response and (**b**) phase response of types A1 and B.

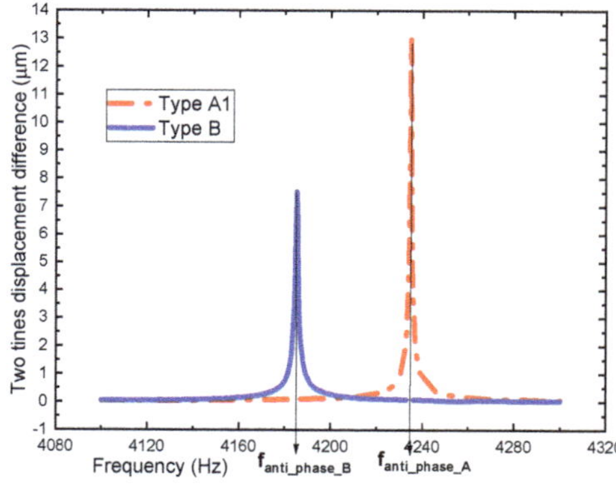

Figure 8. Differential displacement of two tines of types A1 and B.

4.3. Numerical and Theoretical Comparisons

The differential displacement of two tines can be calculated through the analytical expressions of Equations (17) and (19). The theoretical and simulation values of types A1 and B and the error rate are obtained, as listed in Table 4.

Table 4. Comparisons with theoretical and simulation values of types A1 and B.

Type	Type A1			Type B		
Displacement	Theoretical Value	Simulation Value	Error Rate	Theoretical Value	Simulation Value	Error Rate
Two tines' displacement difference (µm)	13.418	12.966	3.49%	7.804	7.532	3.62%

From Table 4, it can be found that the simulation results are consistent with the theoretical values, which verifies the proposed theoretical model. Therefore, the theoretical model can be effective for an anchored leverage mechanism. Meanwhile, the theoretical and simulation values of types A1, A2, and A3 and the error rate are compared in Table 5.

Table 5. Comparison of theoretical and simulation values of types A1, A2 and A3.

Type	Theoretical Value	Simulation Value	Error Rate
Type A1	13.418	12.966	3.49%
Type A2	13.585	13.098	3.72%
Type A3	12.596	12.141	3.75%

Substituting the simulation data in Tables 4 and 5 into Equation (25), the IRMS simulation values of types A1, A2, and A3 are obtained. From Equation (35), the theoretical values of the design can be obtained. The comparison of theoretical and simulation values of types A1, A2, and A3 is shown in Table 6.

Table 6. Comparison of theoretical and simulation values of types A1, A2 and A3.

Type	Theoretical Value	Simulation Value
Type A1	79.10%	72.15%
Type A2	81.33%	73.90%
Type A3	68.06%	61.19%

Table 6 shows that the IRMS of type A1 is lower than that of type A2 and slightly higher than that of type A3 regardless of simulation and theoretical values. It is found that the simulation values of the IRMS are in accord with the theoretical values, although error still exists in these results, which verifies the analysis result of the IRMS in that we proposed that it would increase with increasing α but decrease with increasing β.

4.4. Nonlinear Analysis

Scale factor nonlinearity is one of the main errors of MEMS gyroscopes. The structure error and circuit noises are the main causes of nonlinearity of scale factor. [36–38]. This paper presents nonlinearity caused by large deformation of elastic beam in ideal TFG designs, other potential errors (non-ideal fabrication, capacitive nonlinearity, etc.) are not covered in this paper.

The nonlinear analysis can be used to analyze problems where the stress–strain relationship of the material is nonlinear and check whether the model gives reasonable results. After nonlinear analysis, increasing input force from 0 µN to 7.0×10^4 µN increases displacement of sense-mode frame along y-axis from 0 to 67.916 µm as shown in Figure 9, and a linear reference line are used to highlight the changes in stiffness under large deformation. The nonlinearity of the overall structure cannot be ignored when the deflection of elastic beams exceed a certain limit.

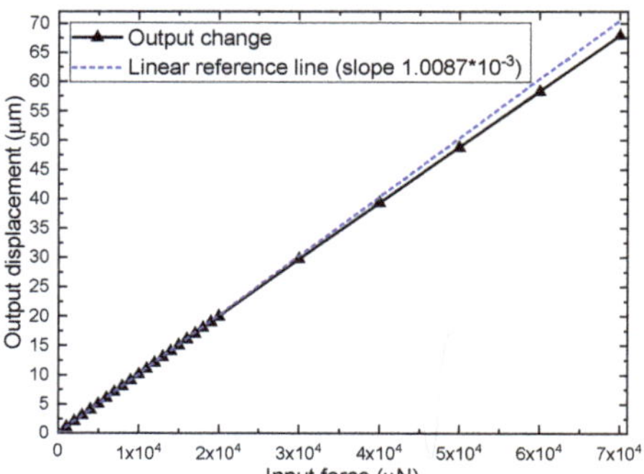

Figure 9. Input force vs. output displacement from 0 μN to 7.0×10^4 μN and linear reference line.

The interval of U-shaped beam in this design is usually relatively small and the width is 20 μm. In addition, solid stoppers should be introduced to improve the shock resistance of the MEMS gyroscope. Therefore, the maximum displacement range of the whole structure is 20 μm. Additionally, to verify the linearity of the TFG in the maximum working range, the deviation from the best-fit line is calculated, as shown in Figure 10. The nonlinearity of the mechanical scale factor is found to be negligible (adjusted $R^2 = 0.999999$), which verified that the design works within a linear interval and the validity of the lumped parameter model proposed.

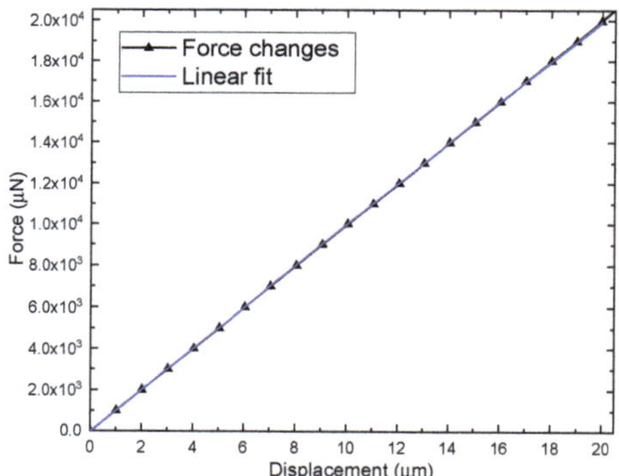

Figure 10. Input force vs. output displacement from 0 μN to 2.0×10^4 μN and linear fit of measured data.

5. Discussion

From the theoretical mechanical analysis of a tuning fork gyroscope with anchored leverage mechanism, it is concluded that the IRMS monotonically increases with increasing α but decreases with increasing β. Actually, the stiffness of the springs associated with SR cannot increase or decrease indefinitely. The strength theory of the structure should be considered at the design stage. On the other hand, stoppers, such as elastic or solid stoppers, should be introduced to improve the shock resistance of the MEMS gyroscope.

It is noteworthy that k_{l1} consists of two U-shaped springs in parallel. This structure has the same effect as the decoupling spring so that it can reach the secondary decoupling effect of quadrature couple in the sense direction.

In future studies, the proposed anchored leverage mechanism of the TFG will be fabricated by a traditional silicon-on-glass and deep reactive ion etching process for experimental verification. In addition, it should be noted that sufficient machining accuracy in the lever mechanism is required to ensure consistency of the stiffness of the left and right levers.

6. Conclusions

In this paper, a novel MEMS tuning fork gyroscope with an anchored leverage mechanism is presented and a new dynamic model is established to investigate the mechanical sensitivity. Moreover, the leverage rate and improvement rate of mechanical sensitivity are analyzed to represent the optimization effect. The theoretical solutions show that the IRMS monotonically increases with increasing α but decreases with increasing β. Three types of gyro structures with different stiffness ratios are designed. In addition, the stiffness of the springs associated with the stiffness ratio is obtained by an FEM simulation. The IRMS of the design can be calculated as 79.10%, 81.33%, and 68.06%. Finally, modal analysis and harmonic response simulation are carried out. FEM simulation demonstrates that the mechanical sensitivities of the design are in accord with theoretical results, verifying the theoretical model. The linearity of design is analyzed, too. Consequently, the anchored leverage mechanism TFG is confirmed to offer a higher displacement output of sense mode, improving the mechanical sensitivity.

Author Contributions: This work was carried out in collaboration among all authors. Z.L. and S.P. investigated and conceived of the work. S.G. provided supervision and guidance in the theoretical analysis. Z.L carried out the theoretical analysis and mechanical design. Z.L. and Y.G. carried out the FEM simulations. L.J. validated the calculation and simulation results. Z.L. wrote the paper. L.J. helped Z.L. improve the quality of the work. H.L. commented on the work.

Funding: This research received no external funding.

Conflicts of Interest: The authors declare no conflict of interest.

References

1. Shkel, A.M. Type I and type II micromachined vibratory gyroscopes. In Proceedings of the IEEE/ION Position, Location, and Navigation Symposium, Coronado, CA, USA, 24–27 April 2006; ION: Coronado, CA, USA, 2006; pp. 586–593.
2. Xia, D.; Yu, C.; Kong, L. The development of micromachined gyroscope structure and circuitry technology. *Sensors* **2014**, *14*, 1394–1473. [CrossRef] [PubMed]
3. Perlmutter, M.; Robin, L. High-performance, low cost inertial MEMS: A market in motion! In Proceedings of the IEEE/ION Position Location and Navigation Symposium (PLANS), Myrtle Beach, SC, USA, 23–26 April 2012; pp. 225–229.
4. Trusov, A.A. *Overview of MEMS Gyroscopes: History, Principles of Operations, Types of Measurements*; University of California: Irvine, CA, USA, 2011.
5. Guan, Y.; Gao, S.; Jin, L.; Cao, L. Design and vibration sensitivity of a MEMS tuning fork gyroscope with anchored coupling mechanism. *Microsyst. Technol.* **2015**, *22*, 247–254. [CrossRef]
6. Guan, Y.; Gao, S.; Liu, H.; Jin, L.; Niu, S. Design and vibration sensitivity analysis of a MEMS tuning fork gyroscope with an anchored diamond coupling mechanism. *Sensors* **2016**, *16*, 468. [CrossRef] [PubMed]
7. Azgin, K.; Temiz, Y.; Akin, T. An SOI-MEMS tuning fork gyroscope with linearly coupled drive mechanism. In Proceedings of the 20th IEEE International Conference on Micro Electro Mechanical Systems (MEMS 2007), Kobe, Japan, 21–25 January 2007; pp. 482–485.
8. Gomez, U.M.K.; Kuhlmann, B.; Classen, J.; Bauer, W.; Lang, C.; Veith, M.; Frey, J.; Grabmaier, F.; Offterdinger, K.; Raab, T.; et al. New surface micromachined angular rate sensor for vehicle stabilizing systems in automotive applications. In Proceedings of the 13th International Conference on Solid-State Sensors, Actuators and Microsystems, Seoul, Korea, 5–9 June 2005; Volume 1.
9. Weinberg, M.S.; Kourepenis, A. Error sources in in-plane silicon tuning-fork MEMS gyroscopes. *J. Microelectromech. Syst.* **2006**, *15*, 479–491. [CrossRef]
10. Sahin, K.; Sahin, E.; Alper, S.E.; Akin, T. A wide-bandwidth and high-sensitivity robust microgyroscope. *J. Micromech. Microeng.* **2009**, *19*, 074004. [CrossRef]

11. Gando, R.K.; Kubo, H.; Masunishi, K.; Tomizawa, Y.; Ogawa, E.; Maeda, S.; Hatakeyama, Y.; Itakura, T.; Ikehashi, T. A catch-and-release drive MEMS gyroscope with enhanced sensitivity by mode-matching. In Proceedings of the 4th IEEE International Symposium on Inertial Sensors and Systems (INERTIAL), Kauai, HI, USA, 27–30 March 2017; pp. 50–53.

12. Xia, D.; Kong, L.; Gao, H. A mode matched triaxial vibratory wheel gyroscope with fully decoupled structure. *Sensors* **2015**, *15*, 28979–29002. [CrossRef] [PubMed]

13. Xu, L.; Li, H.; Ni, Y.; Liu, J.; Huang, L. Frequency tuning of work modes in z-axis dual-mass silicon microgyroscope. *J. Sens.* **2014**, *2014*, 891735. [CrossRef]

14. Bu, F.; Xu, D.; Zhao, H.; Fan, B.; Cheng, M. MEMS gyroscope automatic real-time mode-matching method based on phase-shifted 45 degrees additional force demodulation. *Sensors* **2018**, *18*, 3001. [CrossRef]

15. He, C.; Zhao, Q.; Huang, Q.; Liu, D.; Yang, Z.; Zhang, D.; Yan, G. A MEMS vibratory gyroscope with real-time mode-matching and robust control for the sense-mode. *IEEE Sens. J.* **2015**, *15*, 2069–2077. [CrossRef]

16. Zaman, M.F.; Sharma, A.; Hao, Z.; Ayazi, F. A mode-matched silicon-yaw tuning-fork gyroscope with subdegree-per-hour allan deviation bias instability. *J. Microelectromech. Syst.* **2008**, *17*, 1526–1536. [CrossRef]

17. Sharma, A.; Zaman, M.F.; Ayazi, F. A sub-0.2°/hr bias drift micromechanical silicon gyroscope with automatic CMOS mode-matching. *IEEE J. Solid-State Circuits* **2009**, *44*, 1593–1608. [CrossRef]

18. Xu, L.; Li, H.; Yang, C.; Huang, L. Comparison of three automatic mode-matching methods for silicon micro-gyroscopes based on phase characteristic. *IEEE Sens. J.* **2016**, *16*, 610–619. [CrossRef]

19. Trusov, A.A.; Schofield, A.R.; Shkel, A.M. Micromachined rate gyroscope architecture with ultra-high quality factor and improved mode ordering. *Sens. Actuators A Phys.* **2011**, *165*, 26–34. [CrossRef]

20. Schofield, A.R.; Trusov, A.A.; Shkel, A.M. Versatile vacuum packaging for experimental study of resonant MEMS. In Proceedings of the 23rd IEEE International Conference on Micro Electro Mechanical Systems (MEMS 2010), Hong Kong, China, 24–28 January 2010; pp. 516–519.

21. Trusov, A.A.; Schofield, A.R.; Shkel, A.M. A substrate energy dissipation mechanism in in-phase and anti-phase micromachined z-axis vibratory gyroscopes. *J. Micromech. Microeng.* **2008**, *18*, 095016. [CrossRef]

22. Zotov, S.A.; Simon, B.R.; Prikhodko, I.P.; Trusov, A.A.; Shkel, A.M. Quality factor maximization through dynamic balancing of tuning fork resonator. *Sens. J. IEEE* **2014**, *14*, 2706–2714. [CrossRef]

23. Larkin, K.; Ghommem, M.; Abdelkefi, A. Significance of size dependent and material structure coupling on the characteristics and performance of nanocrystalline micro/nano gyroscopes. *Phys. E Low-Dimens. Syst. Nanostruct.* **2018**, *99*, 169–181. [CrossRef]

24. Zhang, Q.; Feng, L.; Cui, J.; Tang, Y.; Yao, Y. Design of A New Structure Quartz MEMS Gyroscope with High Sensitivity. *IOP Conf. Ser. Mater. Sci. Eng.* **2018**, *382*, 042036. [CrossRef]

25. Jouaneh, M.; Yang, R. Modeling of flexure-hinge type lever mechanisms. *Precis. Eng.* **2003**, *27*, 407–418. [CrossRef]

26. Xu, Q.; Li, Y. Analytical modeling, optimization and testing of a compound bridge-type compliant displacement amplifier. *Mech. Mach. Theory* **2011**, *46*, 183–200. [CrossRef]

27. Iqbal, S.; Malik, A.A.; Shakoor, R.I. Design and analysis of novel micro displacement amplification mechanism actuated by chevron shaped thermal actuators. *Microsyst. Technol.* **2018**, *25*, 861–875. [CrossRef]

28. Iqbal, S.; Shakoor, R.I.; Gilani, H.N.; Abbas, H.; Malik, A.M. Performance Analysis of Microelectromechanical System Based Displacement Amplification Mechanism. *Iran. J. Sci. Technol. Trans. Mech. Eng.* **2018**. [CrossRef]

29. Iqbal, S.; Malik, A.; Shakoor, R.I. Kinematic sensitivity analysis of a novel micro-mechanism for displacement amplification. *Trans. Can. Soc. Mech. Eng.* **2018**, *42*, 436–443. [CrossRef]

30. Iqbal, S.; Shakoor, R.I.; Lai, Y.; Malik, A.M.; Bazaz, S.A. Experimental evaluation of force and amplification factor of three different variants of flexure based micro displacement amplification mechanism. *Microsyst. Technol.* **2019**, *25*, 2889–2906. [CrossRef]

31. Zhang, J.; Su, Y.; Shi, Q.; Qiu, A.P. Microelectromechanical resonant accelerometer designed with a high sensitivity. *Sensors* **2015**, *15*, 30293–30310. [CrossRef] [PubMed]

32. Gao, Y.; Ding, X.; Huang, L.; Li, H. Design and analysis of a novel dual-mass MEMS resonant output gyroscope. *AIP Adv.* **2018**, *8*, 025017. [CrossRef]

33. Gao, Y.; Huang, L.; Ding, X.; Li, H. Design and implementation of a dual-mass MEMS gyroscope with high shock resistance. *Sensors* **2018**, *18*, 1037. [CrossRef] [PubMed]

34. Cao, H.; Li, H. Investigation of a vacuum packaged MEMS gyroscope architecture's temperature robustness. *Int. J. Appl. Electromagn. Mech.* **2013**, *41*, 495–506. [CrossRef]
35. Cao, H.; Li, H.; Kou, Z.; Shi, Y.; Tang, J.; Ma, Z.; Shen, C.; Liu, J. Optimization and Experimentation of Dual-Mass MEMS Gyroscope Quadrature Error Correction Methods. *Sensors* **2016**, *16*, 71. [CrossRef]
36. Yoon, S.W.; Lee, S.; Najafi, K. Vibration-induced errors in MEMS tuning fork gyroscopes. *Sens. Actuators A Phys.* **2012**, *180*, 32–44. [CrossRef]
37. Tang, Q.; Wang, X.; Yang, Q. Scale factor model analysis of MEMS gyroscopes. *Microsyst. Technol.* **2016**, *23*, 1215–1219. [CrossRef]
38. Rezaei Kivi, A.; Azizi, S.; Khalkhali, A. Sensitivity enhancement of a MEMS sensor in nonlinear regime. *Int. J. Mech. Mater. Des.* **2015**, *12*, 337–351. [CrossRef]

Article

Mechanical Behaviors Research and the Structural Design of a Bipolar Electrostatic Actuation Microbeam Resonator

Jingjing Feng [1,4,*], **Cheng Liu** [1,4,*], **Wei Zhang** [2,*], **Jianxin Han** [3] and **Shuying Hao** [1,4]

[1] Tianjin Key Laboratory of the Design and Intelligent Control of the Advanced Mechatronical System, School of Mechanical Engineering, Tianjin University of Technology, Tianjin 300384, China; syhao@tju.edu.cn

[2] Beijing Key Laboratory on Nonlinear Vibrations and Strength of Mechanical Structures, Beijing University of Technology, College of Mechanical Engineering, Beijing 100124, China

[3] Tianjin Key Laboratory of High Speed Cutting and Precision Machining, School of Mechanical Engineering, Tianjin University of Technology and Education, Tianjin 300222, China; hanjianxin@tju.edu.cn

[4] National Demonstration Center for Experimental Mechanical and Electrical Engineering Education, Tianjin University of Technology, Tianjin 300384, China

* Correspondence: jjfeng@tju.edu.cn (J.F.); 163110303@stud.tjut.edu.cn (C.L.); wzhang@bjut.edu.cn (W.Z.); Tel.: +86-226-021-4133 (J.F. & C.L.); +86-106-739-2867 (W.Z.)

Received: 11 February 2019; Accepted: 13 March 2019; Published: 18 March 2019

Abstract: A class of bipolar electrostatically actuated micro-resonators is presented in this paper. Two parametric equations are proposed for changing the microbeam shape of the upper and lower sections. The mechanical properties of a micro-resonator can be enhanced by optimizing the two section parameters. The electrostatic force nonlinearity, neutral surface tension, and neutral surface bending are considered in the model. First, the theoretical results are verified with finite element results from COMSOL Multiphysics simulations. The influence of section variation on the electrostatic force, pull-in behaviors and safe working area of the micro-resonator are studied. Moreover, the impact of residual stress on pull-in voltage is discussed. The multi-scale method (MMS) is used to further study the vibration of the microbeam near equilibrium, and the relationship between the two section parameters of the microbeam under linear vibration was determined. The vibration amplitude and resonance frequency are investigated when the two section parameters satisfy the linear vibration. In order to research dynamic analysis under the case of large amplitude. The Simulink dynamics simulation was used to study the influence of section variation on the response frequency. It is found that electrostatic softening increases as the vibration amplitude increases. If the nonlinearity initially shows hardening behavior, the frequency response will shift from hardening to softening as the amplitude increases. The position of softening-hardening transition point decreases with the increase of residual stress. The relationship between DC voltage, section parameters, and softening-hardening transition points is presented. The accuracy of the results is verified using theoretical, numerical, and finite element methods.

Keywords: parametric equation; finite element methods; Simulink dynamics simulation; softening-hardening transition points

---

1. Introduction

Nowadays, micro electromechanical system (MEMS) devices are attracting attention due to their small size, low weight, and low energy consumption. Among them, microbeams are a popular micro-component and are widely used in MEMS devices, e.g., as a microswitch [1,2], sensor [3,4], or resonator [5–9]. Therefore, designing and optimizing microbeams can improve the mechanical properties of MEMS devices. An electrostatically actuated microbeam resonator is presented in this paper.

Electrostatic actuation is one of the most common actuation methods in microstructures [10]. However, electrostatic actuation also has strong nonlinear characteristics [11]. The pull-in effect is the most critical unstable behavior [12]. When the actuation voltage exceeds a critical value, the microbeam will quickly come into contact with the plate, which can cause damage to the device in severe cases. Therefore, determining the unstable position of the microbeam, increasing the pull-in voltage, and the pull-in position of the resonator to improve its mechanical properties is a key issue in micro-resonator design. Abdel-Rahman et al. [13] presented a mathematical model for electrostatic actuation of microbeam resonators, which takes into account the neutral plane tension and nonlinear electrostatic force. It was found that the maximum dimensionless deflection reached 0.39 near pull-in, which is greater than the result calculated with the traditional method. Mobki, Harried et al. [14] presented two models describing unipolar plate of an actuated microbeam and an actuated bipolar plate microbeam; they found that the bipolar plate actuation model has a higher pull-in voltage. In addition, when the voltage ratio at both ends is 1, the increased gap distance ratio leads to decreased attraction voltage, and vice versa.

In addition to studying the pull-in behavior, vibration of the micro-resonator should also be analyzed. Nonlinearity can lead to multiple solutions, leading to jump instability. The relationship between the system's equivalent natural frequency and DC voltage depends on the strong nonlinearity and electrostatic softening. Moreover, the ideal working condition of the micro-resonator is linear vibration, thus the nonlinearity in the design should be eliminated to the greatest possible extent. Li et al. [15] studied the influence of physical parameters, such as microbeam viscoelasticity and DC voltage, on the system vibration. Finally, the parameters were optimized to determine the relationship between the physical parameters under linear vibration. Han et al. [16] used a spring-mass model to study the vibration of clamped-clamped microbeams and obtained a parameter design diagram relating the DC voltage and initial gap distance under linear conditions. The optimal design voltage under linear conditions was derived. In addition, nonlinearity can be utilized; Li et al. [17] used the modal coupling of a clamped-clamped microbeam to adjust the pull-in voltage and resonant frequency. Based on the coupled vibration behavior between the antisymmetric and symmetric modes, Hu et al. [18] presented a theoretical method for suppressing the midpoint displacement and reducing deformation. However, most early studies were based on rectangular microbeams and focused on optimizing their mechanical properties by adjusting their physical parameters.

As research has deepened, many scholars have found that the mechanical properties of a micro-resonator can be improved by adjusting the microbeam geometry when using the same raw materials. Joglekar and Trivedi [19,20] studied a class of clamped-clamped microbeams and proposed a parametric equation that describes smooth changes along the width of the microbeams. The system mechanical properties are improved by optimizing parameters in the relevant equation, and the results from the method were verified in several cases. On the basis of [19,20], Zhang et al. [21] discussed the influence of the optimized shape on the dynamic response of the microbeam. Kuang and Chen [22] optimized the mechanical properties of a micro-resonator by adjusting the microbeam thickness and gap distance. The required operating voltage range increased by a factor 6. Other studies focus on active control of micro-resonators; Alsaleem and Younis [23] introduced time-delay feedback into the dynamic control of electrostatically actuated micro-resonators. The influence of time delay displacement and velocity feedback on the system dynamic characteristics is discussed in terms of numerical and experimental results. Both kinds of control can improve the system dynamic behavior, enhance stability, and prevent pull-in under positive feedback gain. This is a very effective method for optimized resonator design.

So far, research on microbeam optimization primarily focuses on adjusting the microbeam size [24], using time delay feedback for active control [25,26], and optimizing the shape of the microbeam [27]. In this study, a micro-resonator was optimized by adjusting the microbeam shape, and the number of parametric equations was increased compared to models presented by predecessors. Equations are defined for the upper and lower sections of the microbeam such that the thickness changes uniformly along its length. Each equation has an independent parameter that can be used to adjust variations in the microbeam thickness. The resonator is optimized by choosing the appropriate values of two section parameters. The influence of section on the electrostatic force, small amplitude vibration, and dynamic behavior with large amplitude actuation in bipolar plates are investigated theoretically and with finite element simulations.

The structure of this paper is as follows. The mathematical model (partial differential equation) for electrostatic actuation of a microbeam resonator considering neutral stretching and bending is established in Section 2,. Two parametric equations are proposed for adjusting variations in the microbeam thickness. The Galerkin method is used to simplify the original vibration equation into an ordinary differential equation. In Section 3, the finite element method is used to simulate the electrostatic force on the system, and the influence of two section parameters on the electrostatic force, static pull-in behavior, and safe working area are discussed. In Section 4, the MMS is used to determine the response of the system under small amplitude vibrations. The relationship between the two section parameters is obtained by taking linear vibration as an optimization condition. Variation in the system's equivalent natural frequency and vibration amplitude are studied given this relationship. COMSOL Multiphysics was used to simulate and verify the natural frequency and electrostatic softening of the optimized micro-resonator. In Section 5, the influence of the section parameters on transitions between softening and hardening of the frequency response curve under large amplitude vibration is discussed. Moreover, the relationship between DC voltage, microbeam thickness, and the transition point is presented. Finally, a summary and conclusions are presented in the last section.

2. Mathematical Model

2.1. Governing Equation

A schematic of an electrically actuated microbeam is shown in Figure 1. The model consists of two fixed plates and a movable microbeam. Microbeams can be divided into six cases corresponding to various parametric equations. Figure 1a shows ordinary rectangular beams. Figure 1b,c shows a microbridge model, where the thickness does not change along beam length. Figure 1d,e shows thickened and thinned beams, respectively. both of which are symmetrical around the neutral plane. Figure 1f shows an irregular beam. A capacitor forms between the microbeam and the plate by applying a DC voltage V_{dc} and an AC voltage $V_{ac}\cos(\Omega t)$ to the plate, thus generating an electrostatic force.

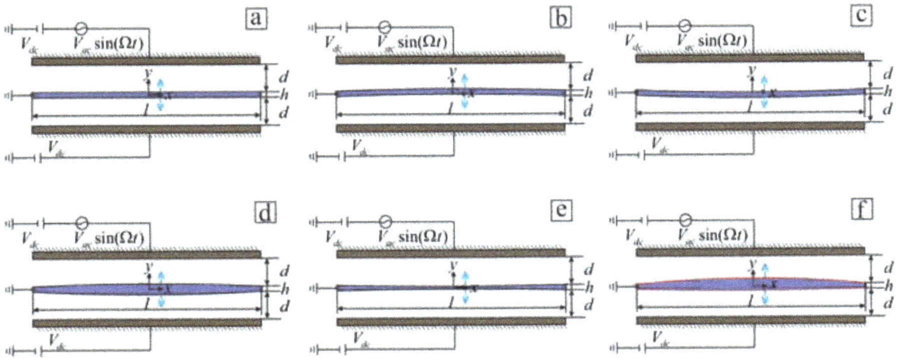

Figure 1. Schematic of an electrically actuated microbeam: (**a**) rectangular beam; (**b**) microbridge bending upward; (**c**) microbridge bending downward; (**d**) thickened microbeam; (**e**) thinned microbeam; (**f**) irregular beam.

Changes in the microbeam shape will produce an uneven electrostatic force between the upper and lower parts of the microbeam. The distance between the microbeam section at the fixed end and electrode plate is d, and the microbeam exhibits lateral motion under the action of electrostatic forces. A model whose thickness varies along microbeam length is considered here, where l and b are the length and the width of the microbeam, respectively. h is the thickness of microbeam at the clamped ends. The cross-sectional area and moment of inertia of the two clamped sides are $A_0 = bh$ and $I_0 = bh^3/12$, respectively. The microbeam thickness is determined by $y_1(x) = \frac{h}{2} + \lambda_1 h \sin\frac{\pi x}{l}$ and $y_2(x) = -\frac{h}{2} + \lambda_2 h \sin\frac{\pi x}{l}$, where λ_1 and λ_2 are section parameters, and the microbeam section curvature changes with λ_1 and λ_2. The cross-sectional area and moment of inertia are $A(x) = A_0(1 + (\lambda_1 - \lambda_2)\sin(\pi x/l))$ and $I(x) = I_0(1 + (\lambda_1 - \lambda_2)\sin(\pi x/l))^3$, respectively. In order to achieve better mechanical properties when both of microbeams consume same amount of material during their fabrication, it is assumed that the volume of the microbeam is constant. The relationship between h and λ_1, λ_2 is found to be $h = \frac{h_0(\pi - 2\lambda_1 + 2\lambda_2)}{\pi}$ by setting the microbeam volume does not change when the shape changes, where h_0 is the microbeam thickness when the beam is rectangular. The capacitance varies in microbeam different positions due to the changes in the gap between the microbeam and electrode plate. The capacitances of the upper and lower plates are

$$C_1 = \frac{\varepsilon_0 \varepsilon_r b l}{d - y_1(x) + \frac{h}{2} - y(x,t)}$$
$$C_2 = \frac{\varepsilon_0 \varepsilon_r b l}{d + y_2(x) + \frac{h}{2} + y(x,t)} \tag{1}$$

where ε_0 is the dielectric constant in free space, ε_r is the relative dielectric permittivity in the gap relative to free space, C_1 is the capacitance between the upper plate and the microbeam, and C_2 is the capacitance between the upper plate and the microbeam.

The total energy stored by two capacitors is

$$W = \frac{1}{2}C_1(V_{dc} + V_{ac})^2 + \frac{1}{2}C_2(V_{dc})^2. \tag{2}$$

The electrostatic force is

$$F = \frac{\partial(\frac{1}{2}C_1(V_{dc} + V_{ac}\cos(\Omega t))^2)}{\partial(d - y_1(x) + \frac{h}{2} - y(x,t))} - \frac{\partial(\frac{1}{2}C_2(V_{dc})^2)}{\partial(d + y_1(x) + \frac{h}{2} + y(x,t))} \tag{3}$$

The equation of motion considers changes in the cross-sectional area and the electrostatic force on the bipolar plate, the equation is therefore rewritten as follows [28]:

$$\frac{\partial^2}{\partial x^2}\left(EI(x)\frac{\partial^2 y(x,t)}{\partial x^2}\right) + \rho A(x)\frac{\partial^2 y(x,t)}{\partial t^2} + c\frac{\partial y(x,t)}{\partial t} = (N$$
$$+ \frac{E}{2l}\int_0^1 A(x)\left(\left(\frac{\partial y(x,t)}{\partial x}\right)^2 + \frac{\partial y(x,t)}{\partial x}\frac{d(y_1(x) + y_2(x))}{dx}\right)dx\right)\left(\frac{\partial^2 y(x,t)}{\partial x^2} + \frac{d^2(y_1(x) + y_2(x))}{2dx^2}\right)$$
$$+ \frac{\varepsilon_0 \varepsilon_r b[V_{dc} + V_{ac}\cos(\Omega t)]^2}{2(d - y_1(x) + \frac{h}{2} - y(x,t))^2} - \frac{\varepsilon_0 \varepsilon_r b V_{dc}^2}{2(d + y_2(x) + \frac{h}{2} + y(x,t))^2} \tag{4}$$

The boundary conditions are

$$y(0,t) = 0, \frac{\partial y}{\partial x}(0,t) = 0, y(l,t) = 0, \frac{\partial y}{\partial x}(l,t) = 0, \tag{5}$$

where c is the damping coefficient, and E and ρ are the elastic modulus and density, respectively.

In Equation (4), N is the residual stress, the integral term is the tension and deformation of neutral plane.

For the convenience of calculation, consider following dimensionless:

$$\widehat{x} = \frac{x}{l}, \widehat{y} = \frac{y}{d}, \widehat{y}_1 = \frac{y_1}{d}, \widehat{y}_2 = \frac{y_2}{d}, \widehat{A}(\widehat{x}) = \frac{A(x)}{A_0}, \widehat{I}(\widehat{x}) = \frac{I(x)}{I_0}, \widehat{t} = \frac{t}{T}, \omega = \frac{\Omega t}{\widehat{t}}. \tag{6}$$

Substituting Equation (6) into (4) and (5) yields the dimensionless bending vibration equation:

$$
\begin{aligned}
&\frac{\partial^2}{\partial \widehat{x}^2}\left(I(\widehat{x})\frac{\partial^2 \widehat{y}(x,t)}{\partial \widehat{x}^2}\right) + A(\widehat{x})\frac{\partial^2 \widehat{y}(x,t)}{\partial \widehat{t}^2} + c'\frac{\partial \widehat{y}(x,t)}{\partial \widehat{t}} - \Big(\widehat{N} \\
&+ \alpha_2\int_0^1 A(\widehat{x})\left(\left(\frac{\partial \widehat{y}(x,t)}{\partial \widehat{x}}\right)^2 + \frac{\partial \widehat{y}(x,t)}{\partial \widehat{x}}\frac{d(\widehat{y}_1(x)+\widehat{y}_2(x))}{d\widehat{x}}\right)d\widehat{x}\Big)\left(\frac{\partial^2 \widehat{y}(x,t)}{\partial \widehat{x}^2}\right. \\
&+ \frac{d^2(\widehat{y}_1(x)+\widehat{y}_2(x))}{2d\widehat{x}^2}\Big) = \alpha_1\Big(\frac{1}{\left(1-\widehat{y}_1(x)+\frac{h}{2d}-\widehat{y}(x,t)\right)^2} - \frac{1}{\left(1+\widehat{y}_2(x)+\frac{h}{2d}+\widehat{y}(x,t)\right)^2}\Big) \\
&+ \frac{2\alpha_1\rho\cos(\widehat{\omega t})}{\left(1-\widehat{y}_1(x)+\frac{h}{2d}-\widehat{y}(x,t)\right)^2} + \frac{\alpha_1\rho^2\cos^2(\widehat{\omega t})}{2\left(1-\widehat{y}_1(x)+\frac{h}{2d}-\widehat{y}(x,t)\right)^2}
\end{aligned}
\tag{7}
$$

The corresponding boundary condition becomes

$$
\widehat{y}(0,\widehat{t}) = 0,\ \frac{\partial \widehat{y}}{\partial \widehat{x}}(0,\widehat{t}) = 0,\ \widehat{y}(1,\widehat{t}) = 0,\ \frac{\partial \widehat{y}}{\partial \widehat{x}}(1,\widehat{t}) = 0,
\tag{8}
$$

where

$$
T = \sqrt{\frac{l^4\rho A_0}{EI_0}};\ c' = \frac{cl^4}{EI_0 T};\ \alpha_1 = \frac{\varepsilon_0\varepsilon_r bl^4 Vdc^2}{2EI_0 d^3};\ \rho = \frac{V_{ac}}{V_{dc}},\ \alpha_2 = 6\left(\frac{d}{h}\right)^2.
$$

In the following figure, the "$\frown$" symbol has been removed from the dimensionless parameters for convenience.

The fringing fields of electrostatic force are not taken into account in the above calculation. In order to observe the accuracy of electrostatic force, the electrostatic force in Equation (7) is compared with the electrostatic force considering edge effect. The results are shown in Figure 2. It can be found that the two results are consistent. Therefore, the infinite plate model is applied for convenience.

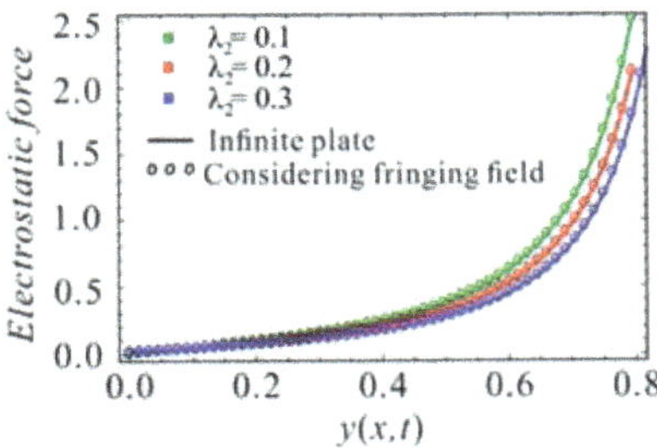

Figure 2. The comparisons of electrostatic forces under different λ_2 at $\lambda_1 = 0$. Solid lines show solution of the infinite plate model and circle show solution considering the fringing fields.

2.2. Galerkin Expansion

The deflection of the microbeam is defined as follows:

$$
y(x,t) = \sum_{i=1}^{\infty} u_i(t)\phi_i(x).
\tag{9}
$$

The boundary conditions are

$$
\phi_i(0) = \phi_i(1) = \phi_i'(0) = \phi_i'(1) = 0,
\tag{10}
$$

$u_i(t)$ is the i-th modal coordinate amplitude, and $\phi_i(x)$ is the i-th order undamped linear orthogonal mode function. According to results from the literature [29], a model with a single degree-of-freedom can sufficiently capture all key nonlinear aspects in the Galerkin approximation when studying the primary resonance. Assuming $y(x,t) = u(t)\phi(x)$, the existing conventional calculation is usually based on a Taylor expansion of the electrostatic force or by multiplying the denominator term of the electrostatic force [30]. The first method is a simple calculation, but the higher order displacement terms will be eliminated after Taylor expansion, which decreases the accuracy of the results. In order to accurately describe the electrostatic force while retaining the nonlinear characteristics of the electrostatic force as much as possible, a second calculation scheme is adopted in this paper. The denominator term is eliminated by multiplying $\left(1 - y_1(x) + \frac{h}{2d} - y(x,t)\right)^2\left(1 + y_2(x) + \frac{h}{2d} + y(x,t)\right)^2$ at vibration equation. Since V_{ac} is much smaller than V_{dc} in the micro-resonator, the following calculation omits the ρ^2 term. Substituting Equation (9) into (7), multiplying by $\phi(x)$, and integrating from x = 0 to 1 yields the following equation:

$$g\ddot{u} + \mu\dot{u} + k_0 + k_1u + k_2u^2 + k_3u^3 + k_4u^4 + k_5u^5 + k_6u^6 + k_7u^7 = 2\alpha_1\rho(v_{ac1} + v_{ac2}u + v_{ac2}u^2)\cos(\omega t), \quad (11)$$

where, $\dot{u} = du/dt$. Expressions for g, μ, k_1, k_2, k_3, k_4, k_5, k_6, k_7, v_{dc1}, v_{dc2}, v_{ac1}, v_{ac2}, and v_{ac3} are shown in Appendix A.

The static equilibrium equation can be written as follows:

$$k_0 + k_1u + k_2u^2 + k_3u^3 + k_4u^4 + k_5u^5 + k_6u^6 + k_7u^7 = 0. \quad (12)$$

Without considering damping and AC disturbances, the following Hamiltonian system corresponding to the dimensionless dynamic equation can be obtained:

$$\begin{cases} \dot{u} = v \\ \dot{v} = -\frac{1}{g}(k_0 + k_1u + k_2u^2 + k_3u^3 + k_4u^4 + k_5u^5 + k_6u^6 + k_7u^7) \end{cases} \cdot \quad (13)$$

The potential energy function is

$$V(\xi) = \int_0^\xi v\,du, \quad (14)$$

and the corresponding Hamiltonian is

$$H(u,v) = V(u) + \frac{1}{2}v^2. \quad (15)$$

One should note here that the maximum lateral displacement of the microbeam is at the midpoint viz., $y_{\max} = \phi(0.5)u \in [-1 - \lambda_2\delta, 1 - \lambda_1\delta]$. The value of the modal function is $\phi(0.5) = 1.59$, thus the range of u is $u \in [\frac{-1-\lambda_2\delta}{1.59}, \frac{1-\lambda_1\delta}{1.59}]$.

3. Static Analysis

First, a static analysis of the system is conducted. The safe region of the micro-resonator can be obtained from static analysis. Only by guaranteeing that the device can operate safely under the selected parameters can the research presented here be meaningful. Here, the accuracy of the electrostatic force obtained from theoretical analysis is verified. All possible microbeam cases are selected. The capacitance the between plates and microbeam was simulated using COMSOL Multiphysics, and the simulation model is shown in Figure 3. A voltage was applied at d on the clamped end of the microbeam while the microbeam was grounded. The entire model was placed in air. Assuming $L = 400$ μm, $b = 45$ μm, $d = 3$ μm, $\rho = 2.33 \times 10^3$ kg/m^3, $E = 165$ Gpa, and $\varepsilon_0 = 8.85 \times 10^{-12}$ F/m, and the thickness of the clamped-clamped ends is $h = 2$ μm. The entire simulation was performed using the steady state solver.

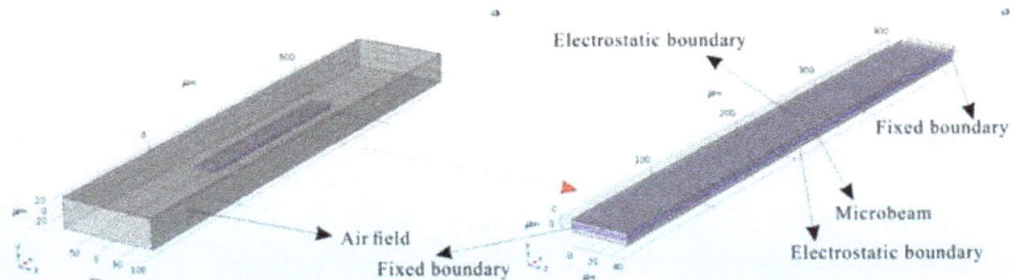

Figure 3. COMSOL simulation model.

The results from two methods are compared in Figure 4. This comparison shows that the two results are consistent. Although there is a slight deviation at larger voltage, the effect is insignificant. At this time, the voltage may have exceeded the pull-in voltage. Thus, high voltage cannot be used during operation. The specific pull-in voltage will be calculated later. Overall, the simulation results verify the correctness of theory.

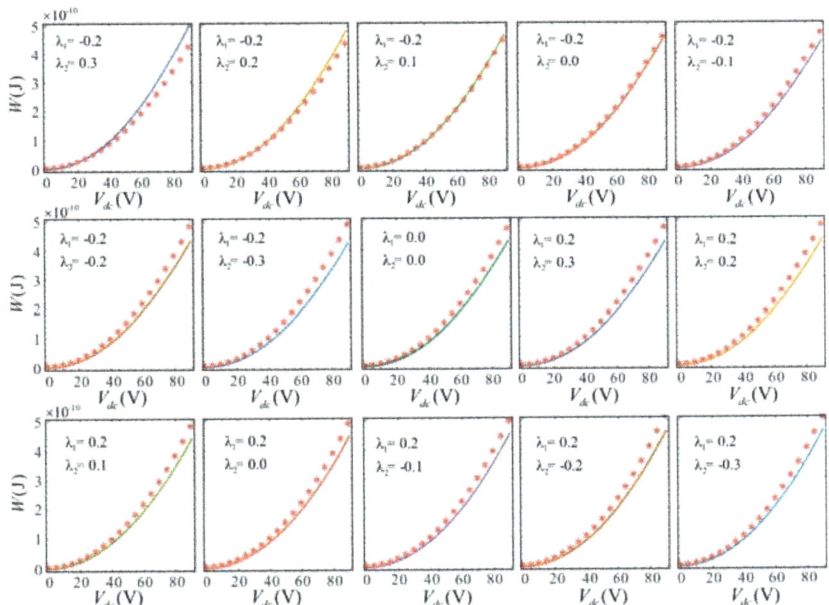

Figure 4. The relationship between the DC voltage and potential energy for different values of section parameters. Lines show the analytical solutions and symbols show the finite element solutions.

The theoretical results shown in Figure 4 are used to determine the influence of section variation on the electrostatic forces. The middle thickness of microbeam becomes thinner and the thickness of clamped-clamped ends becomes thicker when λ_1 is held constant and λ_2 decreases. The system potential can be reduced by decreasing λ_2, as shown in Figure 5. This indicates that the potential energy is larger when the microbeam is thinner at the midpoint for a given voltage due to the greater electrostatic force.

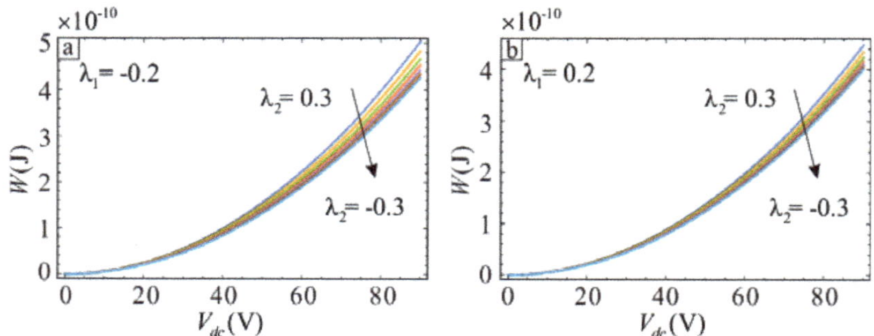

Figure 5. The relationship between the DC voltage and potential energy for different values of section parameters. (**a**) the case of $\lambda_1 = -0.2$; (**b**) the case of $\lambda_1 = 0.2$.

Next, the influence of the section parameters on static pull-in behaviors will be studied. In order to facilitate observation, the special case of a rectangular microbeam is added as a comparison, where the pull-in behavior is indicated with the black curve in Figure 6a,b. The physical parameters of the microbeam are $L = 400$ μm, $b = 45$ μm, $d = 3$ μm, $h_0 = 2$ μm, $E = 1.65 \times 10^{11}$ Pa, and $\rho = 2.33 \times 10^3$ kg/m^3. Figure 6a,b shows the case where the upper section parameters are $\lambda_1 = 0.2$ and $\lambda_1 = -0.2$. The lower section moves gradually from -0.3 to 0.3. At this time, the middle of the microbeam becomes thinner and the ends become thicker. The red curve corresponds to $\lambda_1 = 0.2$ and $\lambda_2 = -0.2$, while the green curve corresponds to $\lambda_1 = -0.2$ and $\lambda_2 = 0.2$. The microbeam is symmetric in the neutral plane in these two cases. Figure 6 shows that the static pull-in position gradually increases as the microbeam becomes thinner, eventually leading to secondary pull-in. Figure 6b shows the case where $\lambda_2 = 0.2$ and $\lambda_2 = 0.3$. Asymmetry of the microbeam will cause the static equilibrium point shift to the side with larger λ_1 and $-\lambda_2$. Greater asymmetry leads to a larger migration distance. When $\lambda_1 = -\lambda_2$, the microbeam is symmetric around the neutral plane and the equilibrium point position is symmetric around the x-axis. If $\lambda_1 = -\lambda_2 > 0$, the midpoint of the microbeam is thicker than at the ends, and vice versa if $\lambda_1 = -\lambda_2 < 0$. Comparing the three curves of red, black, and green in Figure 6a,b shows that the range of pull-in voltage will increase and secondary pull-in phenomenon will be more likely when the midpoint is thinner than the ends.

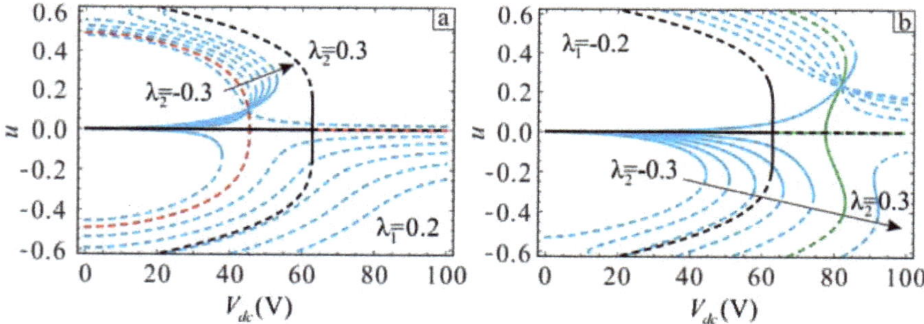

Figure 6. Influence of the section parameters on static pull-in: (**a**) $\lambda_1 = 0.2$ and (**b**) $\lambda_1 = -0.2$. Solid lines show stable solutions and dashed lines show unstable solutions. Blue lines indicate asymmetry, while red and green lines indicate symmetry. The black line indicates the case where the section parameters are $\lambda_1 = \lambda_2 = 0$.

In addition, static displacement jump will occur in the asymmetric microbeam. One can see from the case where $\lambda_2 = 0.3$ in Figure 6b that the static displacement jumps from positive to negative when the DC voltage amplitude increases to $V_{dc} = 87$ V. A comparison of Figure 6a,b shows that the static pull-in voltage mainly depends on the larger side of λ_1 and $-\lambda_2$. As shown in Figure 6a, the static pull-in position of λ_2 from -0.3 to -0.2 varies greatly when $\lambda_1 = 0.2$, but the increasing trend in the static pull-in position decreases significantly when λ_2 ranges from -0.2 to 0.3. This arises because the static suction position at this time is mainly determined by λ_1.

In order to discuss this phenomenon more intuitively, Assume $V_{dc} = 82.5\text{V}$ and draw the potential energy of $\lambda_1 = -0.2$ and $\lambda_2 = 0.2$ to observe the influence of another section variation on system safe area. When $\lambda_1 = -0.2$, the pull-in points of $\lambda_2 = 0.19$, $\lambda_2 = 0.2$, and $\lambda_2 = 0.21$ correspond to green, red, and blue points in Figure 7a, respectively. The lowest barrier energy is significantly lower when compared to the case when $\lambda_2 = 0.2$, $\lambda_2 = 0.21$. Similarly, in Figure 7b, when $\lambda_2 = 0.2$, the phenomenon of $\lambda_1 = -0.21$, $\lambda_1 = -0.2$ and $\lambda_1 = -0.19$ is same as that of Figure 7a. The static pull-in position primarily depends on the larger of λ_1 and $-\lambda_2$. Meanwhile, Figure 7a,b is symmetric around the y-axis, indicating that the corresponding pull-in voltages are equal and the pull-in positions are opposite when the section parameters of two cases are opposite to each other.

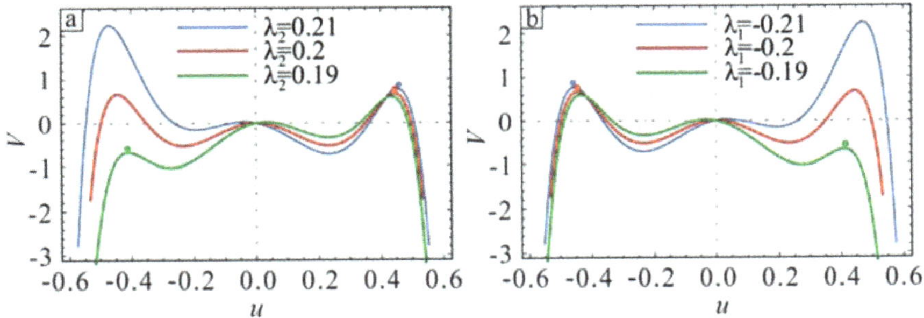

Figure 7. The influence of section parameters on potential energy. (**a**) The influence of λ_2 on static pull-in when $\lambda_1 = -0.2$. (**b**) The influence of λ_1 on static pull-in when $\lambda_2 = 0.2$.

Therefore, reducing λ_1 or increasing λ_2 in the design can improve the working range and stability of the resonator. On the other hand, static pull-in voltage primarily depends on the larger of λ_1 and $-\lambda_2$. When one section of the microbeam is determined, the two fixed ends become thicker and more energy is lost at the ends when the other section is closer to the neutral plane, but the position of pull-in voltage does not change much. Therefore, taking $\lambda_1 = -\lambda_2$ can increase the pull-in voltage.

Residual stress may occur in the process of fabrication or operation of devices and the mechanical properties will be influenced. In order to accurately predict the safe working area of the device, the influence of residual stress on pull-in voltage and secondary pull-in will be discussed below. As shown in Figure 8, the pull-in voltage increases with the increase of residual stress. Figure 8a shows that when $\lambda_1 = 0.2$, the pull-in voltage increases with the increase of λ_2 and before $\lambda_2 < -0.2$, the pull-in voltage increases faster than after $\lambda_2 > -0.2$. This proves once again that the static pull-in voltage primarily depends on the larger of λ_1 and $-\lambda_2$. Figure 8b shows the case of $\lambda_2 = 0.2$, It can be seen that the pull-in voltage decreases as λ_1 increases. Secondary pull-in phenonemon occurs in yellow and blue regions. Each λ_2 corresponds to three voltage values in this region. The dotted line indicates that the voltage on each line corresponds to the position in the static displacement picture. When $\lambda_1 < -0.2$, the secondary pull-in region is close to the lower plate and when $\lambda_1 > -0.2$, the secondary pull-in region is close to upper plate. In addition, the larger the yellow and blue areas, the larger the secondary pull-in range is. Therefore, the increase of residual stress will reduce the range of secondary pull-in area.

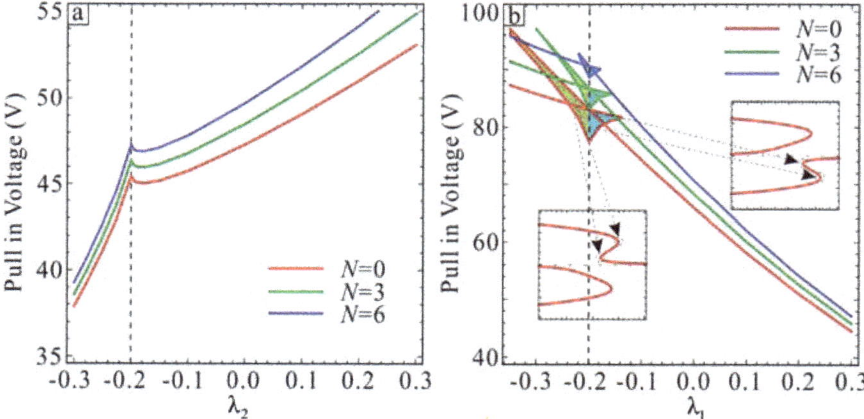

Figure 8. The influence of section parameters on pull-in voltage under different residual stresses. (a) The influence of λ_2 on pull-in voltage when $\lambda_1 = 0.2$. (b) The influence of λ_1 on pull-in voltage when $\lambda_2 = 0.2$.

4. Dynamic Analysis

It is very important to conduct dynamic analysis of MEMS devices. The influence of section variation on the frequency response and bifurcation behavior can be observed through dynamic analysis, so as to optimize micro-resonator. MMS is used in the following to calculate small amplitude vibrations of the micro-resonator in the stable region. Here, we define $u = u_D + u_A$, where u_D is the response under DC voltage and u_A is the response under AC voltage.

First, u_D can be obtained from Equation (12). Substituting $u = u_D + u_A$ into Equation (11), ignoring the nonlinear terms above third order and nonlinear damping in Equation (11), and eliminating the term corresponding to the equilibrium position yields the small vibration equation at the equilibrium position:

$$\ddot{u}_A + c * \dot{u}_A + \omega_n^2 u_A + a_q u_A{}^2 + a_c u_A{}^3 + a_m u_A \ddot{u}_A + a_n u_A{}^2 \ddot{u}_A = f\cos(\omega t). \tag{16}$$

The meaning of each coefficient is shown in Appendix B.

Because V_{ac} is much smaller than V_{dc} in the micro-resonator, we consider the cases $V_{dc} = O(1)$ and $V_{ac} = O(\varepsilon^3)$, where ε is a small dimensionless parameter. Equation (16) can be rewritten as

$$\ddot{u}_A + \varepsilon^2 c * \dot{u}_A + \omega_n^2 u_A + a_q u_A{}^2 + a_c u_A{}^3 + a_m u_A \ddot{u}_A + a u_A{}^2 \ddot{u}_A = \varepsilon^3 f \cos(\omega t). \tag{17}$$

In order to describe the primary resonance accurately, a tuning parameter σ is introduced and defined as

$$\omega = \omega_n + \varepsilon^2 \sigma. \tag{18}$$

The approximate solution to Equation (17) is expressed as follows:

$$u_A(t, \varepsilon) = \varepsilon u_{A1}(T_0, T_1, T_2) + \varepsilon^2 u_{A2}(T_0, T_1, T_2) + \varepsilon^3 u_{A3}(T_0, T_1, T_2), \tag{19}$$

where $T_n = \varepsilon^n t, n = (0, 1, 2)$.

Substituting Equations (18) and (19) into (17) and gathering like powers of ε yields:

$$O(\varepsilon^1) : D_0^2 u_{A1} + \omega_n^2 u_{A1} = 0, \tag{20}$$

$$O(\varepsilon^2) : D_0^2 u_{A2} + \omega_n^2 u_{A2} = -2D_0 D_1 u_{A1} - a_c u_{A1}^2 - a_m D_0^2 u_{A1}, \tag{21}$$

$$O(\varepsilon^3) : \quad D_0^2 u_3 + \omega_n^2 u_3 = -2D_0 D_1 u_{A2} - 2D_0 D_2 u_{A1} - D_1^2 u_{A1} - c^* D_0 u_{A1} - 2a_m u_{A1} D_0 D_1 u_{A1}$$
$$- a_n u_{A1} D_0^2 u_{A2} - a_m u_{A2} D_0^2 u_{A1} - a_n u_{A1}^2 D_0^2 u_{A1} - 2a_c u_1 u_2 - a_c u_1^3 + f\cos(\omega_n T_0 + \sigma T_2) \tag{22}$$

where $D_n = \frac{\partial}{\partial T_n}, n = (0, 1, 2)$.

The general solution to Equation (20) can be written as

$$u_{A1}(T_0, T_1, T_2) = A(T_1, T_2)e^{i\omega_n T_0} + \overline{A}(T_1, T_2)e^{-i\omega_n T_0}. \tag{23}$$

Substituting Equation (23) into (21) yields:

$$D_0^2 u_{A2} + \omega_n^2 u_{A2} = -2i\omega_n \frac{\partial A}{\partial T_1} e^{i\omega_n T_0} - a_q(A^2 e^{2i\omega_n T_0} + A\overline{A}) + a_m \omega_n^2(A^2 e^{2i\omega_n T_0} + A\overline{A}) + cc, \tag{24}$$

where cc denotes the complex conjugate.

To eliminate the secular term, one needs

$$- 2i\omega_n \frac{\partial A}{\partial T_1} e^{i\omega_n T_0} = 0, \tag{25}$$

which indicates that A is only a function of T_2. Thus, Equation (24) becomes

$$D_0^2 u_{A2} + \omega_n^2 u_{A2} = (a_m \omega_n^2 - a_q)(A^2 e^{2i\omega_n T_0} + A\overline{A}) + cc. \tag{26}$$

The solution u_{A2} can be written as follows:

$$u_{A2}(T_0, T_2) = \frac{(a_q - a_m \omega_n^2)A^2}{3\omega_n^2} e^{2i\omega_n T_0} - \frac{(a_q - a_m \omega_n^2)A\overline{A}}{\omega_n^2} + cc. \tag{27}$$

Substituting Equations (23) and (27) into Equation (22) yields the secular terms

$$2i\omega_n \frac{\partial A}{\partial T_1} + c*i\omega A - \frac{(10a_q - a_m \omega_n^2)(a_q - a_m \omega_n^2)A^2 \overline{A}}{3\omega_n^2} + 3(a_c - a_n \omega_n^2)A^2 \overline{A} - \frac{f}{2}e^{i\sigma T_2} = 0. \tag{28}$$

At this point, it is convenient to express A in polar form:

$$A = \frac{1}{2}a(T_2)e^{i\beta(T_2)} + cc. \tag{29}$$

Substituting Equation (29) into Equation (28) and separating the imaginary and real parts yields:

$$\frac{Da}{DT_2} = -\frac{c*}{2}a - \frac{f}{2\omega_n}\sin\varphi, \tag{30}$$

$$a\frac{D\varphi}{DT_2} = \sigma a + a^3\left(\frac{(10a_q - a_m \omega_n^2)(a_q - a_m \omega_n^2)}{24\omega_n^3} - \frac{3(a_c - a_n \omega_n^2)}{8\omega_n}\right) + \frac{f}{2\omega_n}\cos\varphi, \tag{31}$$

where $\varphi = \sigma T_2 - \beta$.

The steady-state response can be obtained by imposing the condition $\frac{Da}{DT_2} = \frac{D\varphi}{DT_2} = 0$. Finally, the frequency response equation can be derived as follows:

$$a^2\left(\left(\frac{c*}{2}\right)^2 + (\sigma + a^2\xi)^2\right) = \left(\frac{f}{2\omega_n}\right)^2, \tag{32}$$

where $\xi = \frac{(10a_q - a_m \omega_n^2)(a_q - a_m \omega_n^2)}{24\omega_n^3} - \frac{3(a_c - a_n \omega_n^2)}{8\omega_n}$.

The peak vibration amplitude and backbone curve are $a_{max} = f/(\mu\omega_n)$ and $\omega = \omega_n - \kappa a_{max}$, respectively. The stability of the periodic solution can be determined using a method found in the literature [29,31].

4.1. Parameter Optimization

In the previous section, the stable region of the micro-resonator is optimized by static analysis, followed by conducting dynamic analysis. Previous research [29] has shown that the section parameter λ changes the softening and hardening behavior of the frequency response. It is noteworthy that there will be a moment of linear vibration in the transformation between softening and hardening, which is an ideal condition for micro-resonators. Therefore, the MEMS resonator is designed based on this condition. Assume $d = 2$ μm. First, one finds by calculation that the microbeam vibration is linear and rectangular when $V_{dc} = 23.95$ V. The sectional parameters λ_1 and λ_2 under linear vibration are shown in Figure 9.

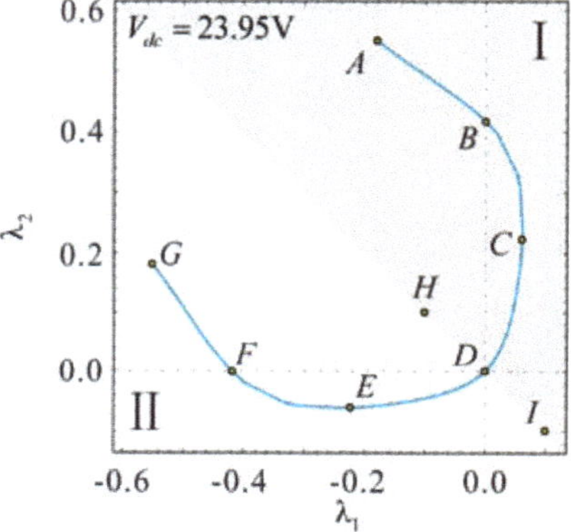

Figure 9. The relationship between λ_1 and λ_2 under linear vibration when $V_{dc} = 23.95$ V.

The coordinates of these special points listed in Table 1 are shown in Figure 9. The frequency response in nine cases was studied using MMS. In order to verify the above theoretical results, the long-time integral method (LTI) is used to calculate Equation (11) and the numerical solution is obtained. First, the frequency responses of D, H, and I are shown in Figure 10. Adjusting the AC voltage amplitude to a frequency response maximum amplitude of 0.1 shows that the vibration inside the curve exhibits hardening while the outer side exhibits softening. The point D corresponds to linear vibration in the special case of $\lambda_1 = 0$ and $\lambda_2 = 0$.

Table 1. Section parameter corresponding to nine special points.

Special Points	λ_1	λ_2
A	−0.18	0.55
B	0	0.419
C	0.061	0.225
D	0	0
E	−0.225	−0.061
F	−0.419	0
G	−0.55	0.18
H	−0.1	0.1
I	0.1	−0.1

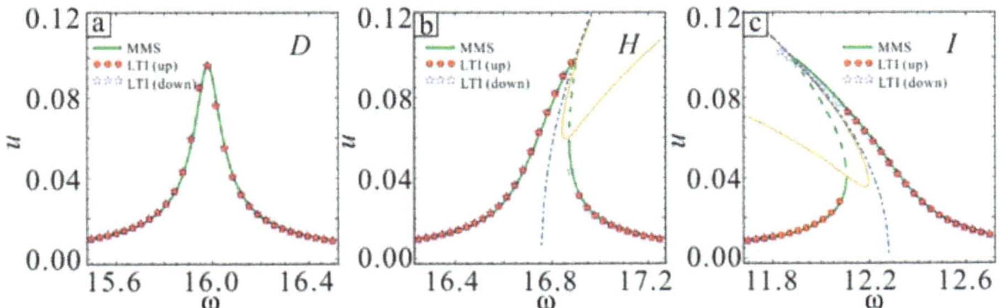

Figure 10. Frequency response curves for D (**a**), H (**b**), and I (**c**). The solid line shows the theoretically stable solution and the dotted line shows the theoretically unstable solution.

Next, points A, B, C, and F in Figure 9 are analyzed to determine the influence of section variation on static pull-in so as to increase the working range of the micro-resonator, where point B and F are symmetric about $y = x$. Figure 11a shows that the maximum static suction voltage is obtained when $\lambda_1 = -0.18$ and $\lambda_2 = 0.55$. This also verifies the conclusion in Section 3, namely that the pull-in voltage is determined by the larger value out of λ_1 and $-\lambda_2$. The static pull-in curve is symmetric around $x = 0$ when the point in Figure 7 is symmetric about $y = x$, where the static pull-in voltage is equal and pull-in position is the opposite, as shown in Figure 11b.

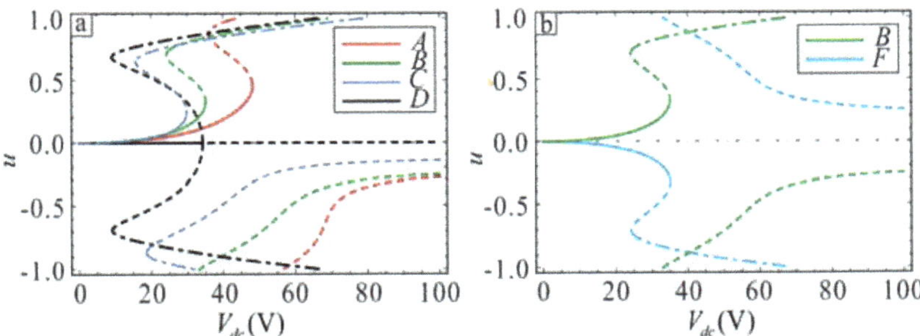

Figure 11. The relationship between the DC voltage and the static equilibrium point. (**a**) Static equilibrium point for A, B, C, and D. (**b**) Static equilibrium point for B and F cases. Solid lines represent stable solutions and dashed lines represent unstable solutions. The dotted line shows the stable solution without physical meaning.

The equivalent frequency and amplitude of the microbeam in different conditions are also important for studying the micro-resonator. Figure 12 shows the frequency response curves in different conditions. It is found that the equivalent frequencies and amplitudes of various points in Figure 9 are different. In addition, the points that are symmetric about $y = x$ in Figure 12 have the same equivalent frequency and different amplitudes. The vibration amplitude in region I is relatively large.

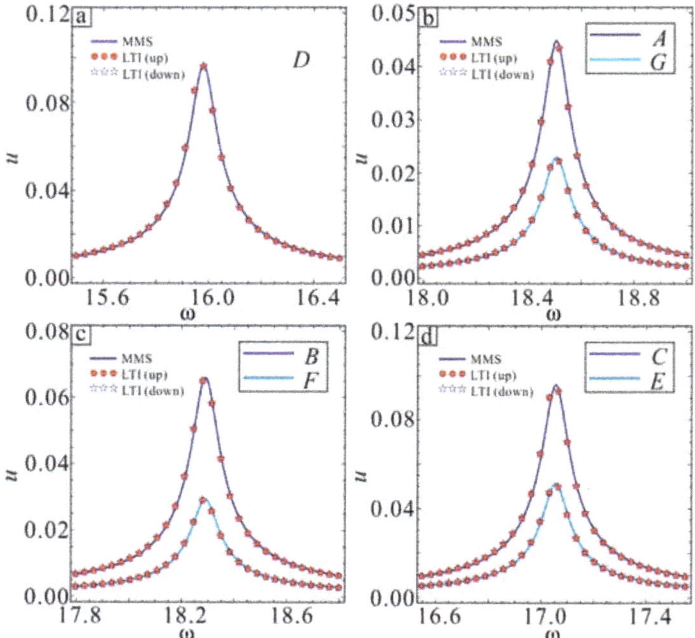

Figure 12. Frequency response curves for A, B, C, D, E and F cases: (**a**) the case of D; (**b**) the cases of A and G; (**c**) the cases of B and G; (**d**) the cases of C and E.

The variation in the resonance frequency and amplitude with the sectional parameters λ_1 and λ_2 are studied based on the results in Figure 9. Figure 13a shows that the resonance frequency is largest at point A, and the resonance frequency first decreases and then increases from point A to point B. From point B to point D, the resonance frequency decreases rapidly, reaching point $\lambda_1 = 0$, $\lambda_2 = 0$, and the resonance frequency is smallest. The trends of G, E, and F are the same as those of A, B, and C, and the corresponding equivalent frequency is equal at the point where $y = x$ is symmetric. As shown in Figure 13b, the resonance amplitude increases from point A to point C. For the point corresponding to the $y = x$ symmetry in Figure 9, the amplitude is larger when λ_1 is larger. The maximum amplitude appears between C and D.

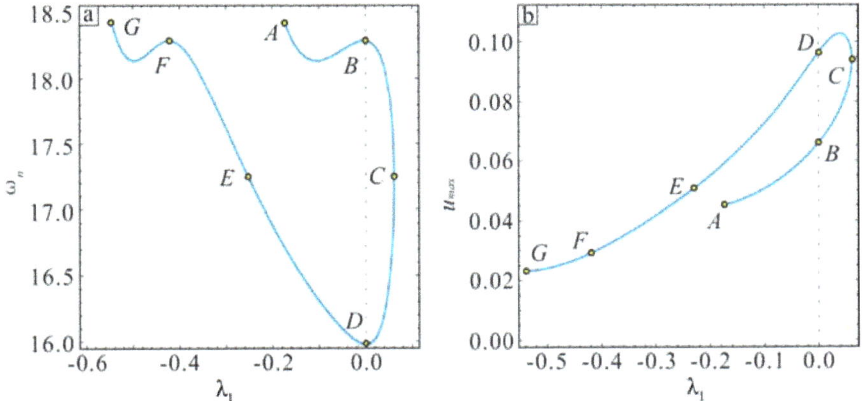

Figure 13. Influence of the section parameters on system vibration characteristics under linear vibration.
(a) Influence of the cross-section parameters on the equivalent natural frequencies under linear vibration.
(b) Influence of the section parameters on vibration amplitude under linear vibration.

4.2. Dynamic Analysis with Large Amplitude

Since traditional MMS is only suitable for small vibrations, the traditional MMS will produce incorrect results when vibration amplitude exceeds a certain value [32]. In order to study the mechanical properties of the system when the vibration exceeds a certain amplitude, a new multi-scale method (NMMS) is introduced by combining the homotropy idea with the multi-scale method [33]. Equation (11) is solved using a Simulink dynamics simulation in order to verify the accuracy of the results.

It is assumed that the final vibration frequency is equal to the excitation frequency. A scaling parameters ε can be used to convert Equation (11) into the following form

$$
\begin{aligned}
\ddot{u} + \omega^2 u = \ & \varepsilon \Big\{ -\tfrac{c'C_{22}}{B_{22}}\dot{u} + \big(\omega^2 - \tfrac{k_1}{B_{22}}\big)u - \tfrac{k_0}{B_{22}} - \tfrac{k_2}{B_{22}}u^2 - \tfrac{k_3}{B_{22}}u^3 - \tfrac{k_4}{B_{22}}u^4 - \tfrac{k_5}{B_{22}}u^5 - \tfrac{k_6}{B_{22}}u^6 - \tfrac{k_7}{B_{22}}u^7 \\
& + \tfrac{2B_{12}-2B_{21}}{B_{22}}u\ddot{u} + \tfrac{4B_{11}-B_{20}-B_{02}}{B_{22}}u^2\ddot{u} + \tfrac{2B_{10}-2B_{01}}{B_{22}}u^3\ddot{u} - \tfrac{B_{00}}{B_{22}}u^4\ddot{u} \\
& + \tfrac{2\alpha_1\rho}{B_{22}}\big(v_{ac1} + v_{ac2}u + v_{ac2}u^2\big)\cos(\omega t) \Big\}
\end{aligned}
\tag{33}
$$

The approximate solution to Equation (33) is expressed as follows:

$$
u(t,\varepsilon) = u_0(T_0, T_1, T_2) + \varepsilon u_1(T_0, T_1, T_2) + \varepsilon^2 u_2(T_0, T_1, T_2) + \cdots .
\tag{34}
$$

Substituting Equation (34) into (33) and equating coefficients of like powers of ε yields the following equations:

$$
O(\varepsilon^0) : D_0^2 u_0 + \omega^2 u_0 = 0.
\tag{35}
$$

$$
\begin{aligned}
O(\varepsilon^1) : \ & D_0^2 u_1 + \omega^2 u_1 = -2D_0 D_1 u_1 - \tfrac{c'C_{22}}{B_{22}}D_0 u + \big(\omega^2 - \tfrac{k_1}{B_{22}}\big)u_0 - \tfrac{k_0}{B_{22}} - \tfrac{k_2}{B_{22}}u_0^2 - \tfrac{k_3}{B_{22}}u_0^3 \\
& - \tfrac{k_4}{B_{22}}u_0^4 - \tfrac{k_5}{B_{22}}u_0^5 - \tfrac{k_6}{B_{22}}u_0^6 - \tfrac{k_7}{B_{22}}u_0^7 + \tfrac{2B_{12}-2B_{21}}{B_{22}}u_0 D_0^2 u_0 + \tfrac{4B_{11}-B_{20}-B_{02}}{B_{22}}u_0^2 D_0^2 u_0 \\
& + \tfrac{2B_{10}-2B_{01}}{B_{22}}u_0^3 D_0^2 u_0 - \tfrac{B_{00}}{B_{22}}u_0^4 D_0^2 u_0 + \tfrac{2\alpha_1\rho}{B_{22}}\big(v_{ac1} + v_{ac2}u_0 + v_{ac2}u_0^2\big)\cos(\omega t)
\end{aligned}
\tag{36}
$$

where: $D_n = \tfrac{\partial}{\partial T_n}$, $n = (0, 1, 2)$.

The general solution to Equation (35) can be written in the following form:

$$
u_0 = A(T_1)\cos(\omega T_0 + \beta(T_1)),
\tag{37}
$$

where $A(T_1)$ is vibration amplitude and $\beta(T_1)$ is the vibration phase.

The secular term derived from MMS is expressed as:

$$\int_0^{2\pi}\Bigg(-2D_0D_1u_1 - \frac{c'C_{22}}{B_{22}}D_0u + (\omega^2 - \frac{k_1}{B_{22}})u_0 - \frac{k_0}{B_{22}} - \frac{k_2}{B_{22}}u_0{}^2 - \frac{k_3}{B_{22}}u_0{}^3 - \frac{k_4}{B_{22}}u_0{}^4 - \frac{k_5}{B_{22}}u_0{}^5 - \frac{k_6}{B_{22}}u_0{}^6$$
$$- \frac{k_7}{B_{22}}u_0{}^7 + \frac{2B_{12}-2B_{21}}{B_{22}}u_0D_0^2u_0 + \frac{4B_{11}-B_{20}-B_{02}}{B_{22}}u_0{}^2D_0^2u_0 + \frac{2B_{10}-2B_{01}}{B_{22}}u_0{}^3D_0^2u_0 - \frac{B_{00}}{B_{22}}u_0{}^4D_0^2u_0$$
$$+ \frac{2\alpha_1\rho}{B_{22}}(v_{ac1} + v_{ac2}u_0 + v_{ac2}u_0{}^2)\cos(\omega t)\Bigg)\exp(-i\varphi)\mathrm{d}\varphi \tag{38}$$

where $\varphi = \omega T_0 + \beta(T_1)$.

Substituting Equation (37) into Equation (38) and separating the imaginary and real parts yields:

$$\begin{cases}\dfrac{da}{dT_1} = \dfrac{ac'C_{22}}{2B_{22}} - \dfrac{v_{ac1}\alpha_1\rho}{B_{22}\omega}\sin(\beta) - \dfrac{a^2v_{ac3}\alpha_1\rho}{4B_{22}\omega}\sin(\beta)\\[2ex] a\dfrac{d\beta}{dT_1} = -\dfrac{ak_1}{2B_{22}\omega} + \dfrac{3a^3k_3}{8B_{22}\omega} + \dfrac{5a^5k_5}{16B_{22}\omega} + \dfrac{35a^7k_7}{128B_{22}\omega} - \dfrac{a\omega}{2} - \dfrac{3a^3B_{20}\omega}{8B_{22}} - \dfrac{3a^3B_{02}\omega}{8B_{22}}\\[2ex] \qquad + \dfrac{3a^3B_{11}\omega}{2B_{22}} - \dfrac{5a^5B_{00}\omega}{16B_{22}} - \dfrac{v_{ac1}\alpha_1\rho}{B_{22}\omega}\cos(\beta) - \dfrac{3a^2v_{ac3}\alpha_1\rho}{4B_{22}\omega}\cos(\beta)\end{cases} \tag{39}$$

The steady-state response can be obtained by applying the condition $Da/DT_1 = D\varphi/DT = 0$. The ultimate frequency response can be deduced as follows:

$$4a^2c'^2C_{22}^2\omega^2/(4vac_1 + a^2vac_3)^2\alpha_1^2\rho^2 + (64ak_1 + 48a^3k_3 + 40a^5k_5 + 35a^7k_7 - 48a^3B_{20}\omega^2$$
$$-48a^3B_{02}\omega^2 + 192a^5B_{11}\omega^2 - 40a^5B_{00}\omega^2 - 60aB_{22}\omega^2)^2/1024(4vac_1 + 3a^2vac_3)^2\alpha_1^2\rho^2 = 1 \tag{40}$$

The micro-resonator vibration amplitude increases as the AC voltage amplitude increases. The physical parameters of the microbeam are $L = 400\ \mu\mathrm{m}$, $b = 45\ \mu\mathrm{m}$, $d = 3\ \mu\mathrm{m}$, $h_0 = 2\ \mu\mathrm{m}$, $V_{dc} = 60\ \mathrm{V}$, $E = 1.65 \times 10^{11}\ \mathrm{Pa}$, and $\rho = 2.33 \times 10^3\ \mathrm{kg/m^3}$. Select the upper section parameters $\lambda_1 = -0.2$ and $\lambda_2 = 0.2$. The results from MMS, NMMS, and the Simulink dynamics simulation are compared in Figure 14, and the results from the three methods are found to be consistent when $V_{ac} = 0.05\ \mathrm{V}$. However, the MMS results begins to exhibit errors when $V_{ac} = 0.1\ \mathrm{V}$ as the vibration amplitude increases, while the NMMS results are still consistent with the simulation results. The system vibration appears to transition from hardening to softening when $V_{ac} = 0.2\mathrm{V}$. The new behavior can be obtained with NMMS. It is worth noting that the frequency response curve is composed of high and low energy branches when the vibration amplitude is small. As the amplitude increases, the two branches gradually converge. Since the main purpose of this chapter is to discuss the nonlinear softening-hardening transition behavior, it will not be discussed in further detail here. A detailed discussion can be found in literature [28].

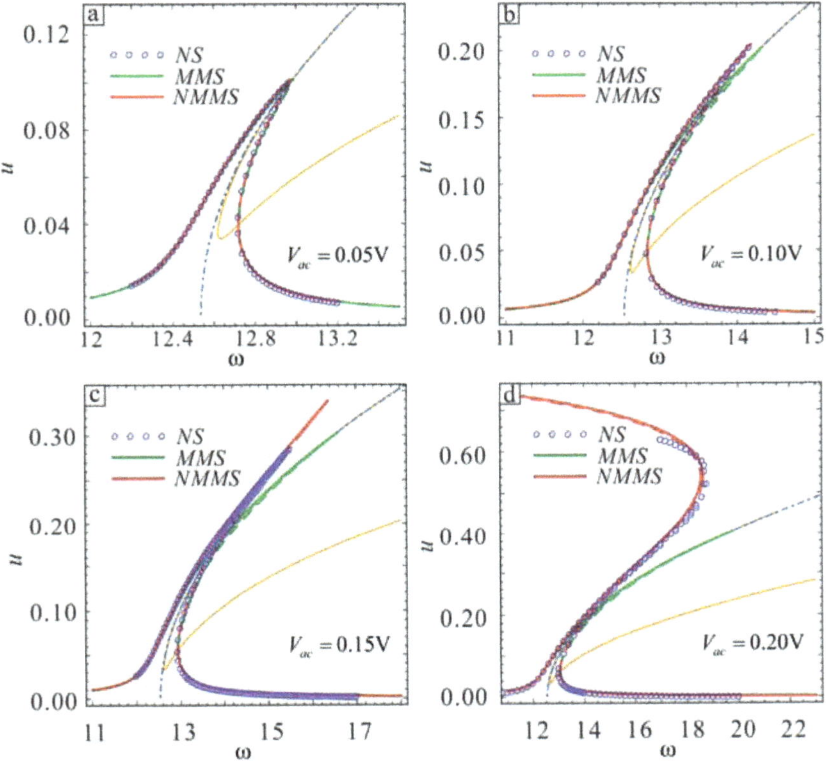

Figure 14. Influence of the AC voltage on the frequency response curve when (**a**) $V_{ac} = 0.05$ V, (**b**) $V_{ac} = 0.1$ V, (**c**) $V_{ac} = 0.15$ V, and (**d**) $V_{ac} = 0.20$ V. The green straight line shows the theoretically stable solution obtained with MMS. The green dotted line shows the theoretically unstable solution obtained with MMS. The red straight line shows a theoretically stable solution obtained with NMMS. The red dotted line shows the theoretically unstable solution obtained with NMMS. Circles show the numerical solution obtained with the Simulink dynamics simulation (NS).

Two kinds of motion can occur at the transition point: dynamic pull-in and jump to a higher stable branch. The energy output is higher in the second case. Changes in the softening-hardening transition point can be observed as the vibration amplitude increases from $V_{ac} = 0.2$ V to $V_{ac} = 0.25$ V, $V_{ac} = 0.3$ V, and $V_{ac} = 0.35$ V, and the corresponding frequency response curve can be drawn. One can see from Figure 15 that the displacements corresponding to nonlinear softening-hardening transition points in the four cases are almost the same. It can be inferred that the position of the transition point does not change as the vibration amplitude increases or the change is negligible at the transition point.

According to previous studies, the location of the transition point is related to the microbeam thickness and DC voltage. The location of the transition point can be obtained by calculating $du/d\omega = 0$, yielding the relationship between the thickness, DC voltage, and transition point. The transition point can only be obtained numerically because the resulting equation is more complicated. Curve fitting is used here to determine the transition point. There are two transition points in each frequency response curve. Figure 16 shows that the amplitudes corresponding to the two transition points are nearly equal. Therefore, only one curve can be used for the two positions. As can be seen from Figure 14, the position of softening-hardening transition point decreases with the increase of residual stress. Figure 16a shows the relationship between the thickness and the transition point position, and Figure 16b shows the relationship between the DC voltage and the transition point position. The straight line shows the solution drawn with the fitted equation, and the circle shows the numerical solution obtained by the Newton iteration. The green and blue circles show the left and right transition points, respectively. A comprehensive analysis of the two curves yields the following equation:

$$u = 0.499 - 0.0057V_{dc} - 1.1613(\lambda_1 - \lambda_2) - 0.4846(\lambda_1 - \lambda_2)^2. \tag{41}$$

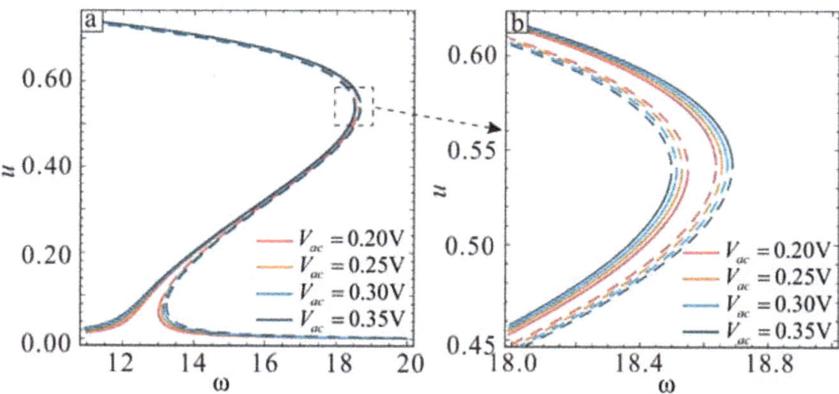

Figure 15. Influence of the AC voltage on the frequency response curve. The straight line shows the stable solution and the dotted line shows the unstable solution.

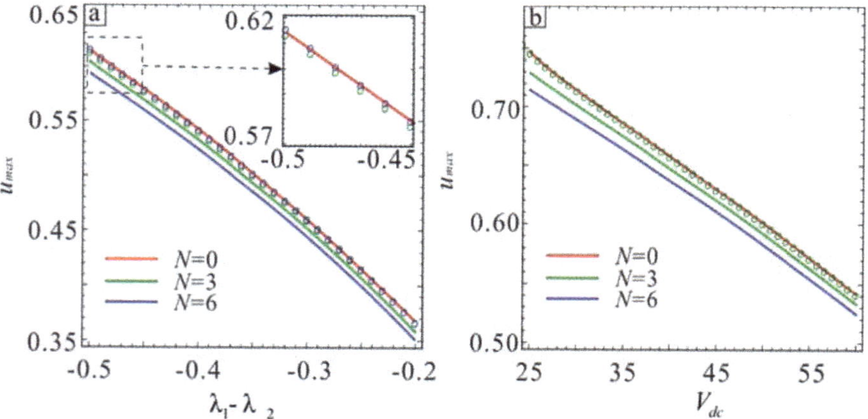

Figure 16. The relationship between the transition points and physical parameters of the system under different residual stresses. (**a**) The relationship between the transition points and $\lambda_1 - \lambda_2$. (**b**) The relationship between the transition points and DC voltage. The straight line shows the result obtained with Equation (41). Circles show the exact position of the transition point obtained by the Equation (40).

Next, the relationship between the DC voltage, microbeam thickness, and transition point position is verified. Table 2 shows the case where the voltages are $V_{dc} = 30$ V, $V_{dc} = 40$ V, $V_{dc} = 50$ V, and $V_{dc} = 60$ V. Each case considers four different thicknesses to account for the breadth of the data. u_e is the transition point position obtained from empirical equation, $u_{n\ l}$ and $u_{n\ r}$ are the positions of left and right transition points obtained by setting $du/d\omega = 0$, respectively, $u_{n\ l}$ is the left transition point, and $u_{n\ r}$ is the right transition point. The "/" symbol indicates no transition point, i.e., there is no *softening-hardening* transition. Table 2 shows that the error between the transition point position and the actual transition point obtained from Equation (41) is small, which verifies the accuracy of the equation.

Table 2. The comparison of transition points obtained with two methods.

V_{dc} (V)	$\lambda_1-\lambda_2$	u_e	$u_{n\,l}$	$u_{n\,r}$	Error
60	-0.6	0.6793	0.6882	0.6905	$-1.29\%/-1.62\%$
	-0.5	0.6165	0.6133	0.6156	0.52%/0.15%
	-0.4	0.5440	0.5306	0.5326	2.53%/2.14%
	-0.3	0.4618	0.4364	0.4373	5.82%/5.60%
50	-0.5	0.6735	0.6729	0.6757	$0.09\%/-0.33\%$
	-0.4	0.6010	0.5927	0.5958	1.40%/0.87%
	-0.3	0.5188	0.5045	0.5070	2.83%/2.33%
	-0.2	0.4269	0.4046	0.4049	5.51%/5.43%
40	-0.4	0.6580	0.6546	0.6585	$0.52\%/-0.08\%$
	-0.3	0.5758	0.5686	0.5722	1.27%/0.63%
	-0.2	0.4839	0.4742	0.4767	2.05%/1.51%
	-0.1	0.3823	0.3682	0.3674	3.83%/4.06%
30	-0.3	0.6328	0.6349	0.6412	$-0.33\%/-1.31\%$
	-0.2	0.5409	0.5386	0.5439	0.43%/1.12%
	-0.1	0.4393	0.4370	0.4390	0.53%/0.07%
	0.0	0.3280	0.3254	0.3211	0.80%/2.15%

5. Finite Element Verification

5.1. Equivalent Natural Frequency Simulation

The microbeam resonator was explored and optimized using theoretical and numerical analysis. The finite element method will be used to simulate the optimization results using COMSOL Multiphysics software.

In COMSOL, the "solid mechanics" interface and "electrostatic" interface are combined with a dynamic grid function. First, the steady state of the model and the eigenfrequency in the steady state are studied. Here, finite element simulations are performed for the aforementioned seven cases, and the calculated simulation results are shown in Figure 17. One can see that the theoretical and simulation results are consistent, and the pull-in voltage at A, B, C, D, and G in Figure 17 are consistent with those in Figure 11.

The seven cases correspond to linear vibration when $V_{dc} = 23.95$ V. The equivalent theoretical frequency results are compared with the finite element equivalent frequency results when $V_{dc} = 23.95$ V in Table 3. It is found that the maximum error is only 5.49%, which shows that the theoretical and numerical methods are consistent.

Table 3. Comparison of the finite element simulation and theoretical results.

	A/G	B/F	C/E	D	H	I
Theoretical frequency	18.507	18.289	17.050	15.973	16.752	12.269
Simulation frequency	17.876	18.530	16.371	15.354	16.621	11.631
Error	3.51%	-1.30%	4.15%	4.03%	0.79%	5.49%

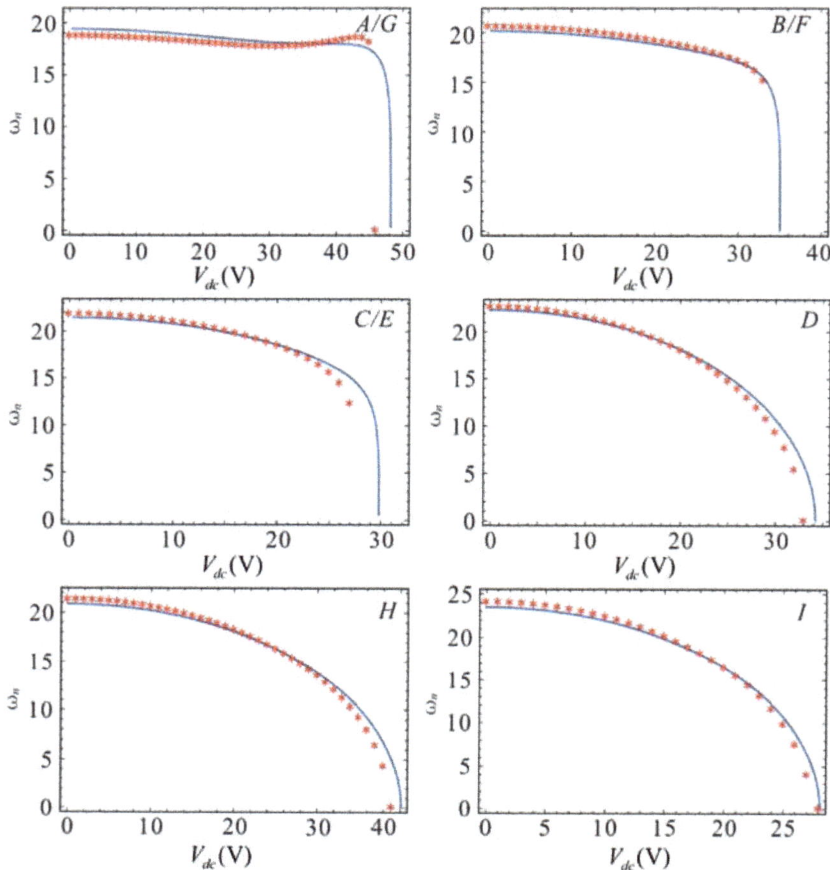

Figure 17. Electrostatic force softening effect in different cases. Lines show the analytical solutions and pentagons show the finite element solutions.

5.2. Equivalent Natural Frequency Simulation

The simulated equivalent natural frequency contains some error, although the error between the finite element solution and the theoretical solution is relatively small. More simulated frequency response curves are presented in this section. The physical parameters of the microbeam are $L = 400$ μm, $b = 45$ μm, $d = 3$ μm, $h_0 = 2$ μm, $V_{dc} = 60$ V, $E = 1.65 \times 10^{11}$ Pa, and $\rho = 2.33 \times 10^3$ kg/m^3. The simulated sectional parameters were $\lambda_1 = -0.2$ and $\lambda_2 = 0.2$. Figure 16 shows a comparison between the simulation and theoretical results.

One can see that the amplitude response is linear when the amplitude is small, while hardening appears at higher amplitude. The red asterisk shows the results obtained from the frequency response analysis in COMSOL. One can see that the simulated amplitude is consistent with the theoretical amplitude when the AC voltage varies. When $V_{ac} = 0.02$V, the simulated amplitude is 0.0409 and the theoretical amplitude is 0.0406, yielding a relative error of -0.73%. When $V_{ac} = 0.1$ V, the simulated amplitude is 0.204 and the theoretical amplitude is 0.203, yielding a relative error of 0.49%. However, because the frequency response analysis solver is a linear solver, the results do not exhibit hardening or softening. Each frequency is selected separately in order to further simulate the nonlinearity. After determining the frequency at a certain point, the corresponding amplitude is obtained with the transient solver. This step was repeated with different frequencies in order to determine the corresponding amplitude. This result is shown in blue circles in Figure 18. One can see that the finite element solution exhibits hardening. Moreover, the simulated natural frequency in the frequency domain is the same as the natural frequency in the time domain. The two simulation results are very consistent when the amplitude is small, but the maximum vibration amplitude from the time domain is lower than the other two results as the amplitude increases. This phenomenon may arise because the time interval corresponding to the highest point position is too short in the time domain analysis, thus the displacement does not point near the peak.

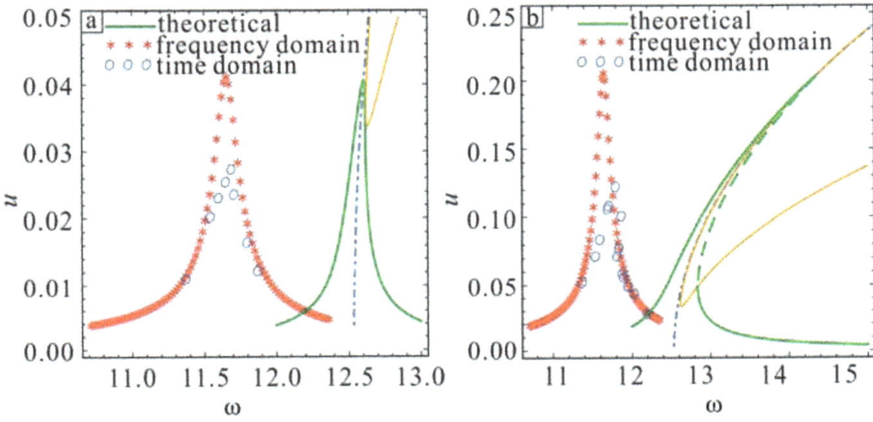

Figure 18. Comparison of frequency response curves when (**a**) $V_{ac} = 0.02$ V and (**b**) $V_{ac} = 0.1$ V. The line shows the analytical solution, asterisks show the finite element solution obtained with the frequency domain solver, and circles show the finite element solution obtained with the time domain solver.

6. Conclusions

Static and dynamic analyses of electrostatically-driven microbeam resonators are presented in this paper. Two parametric equations describing control over the upper and lower sections of microbeam were introduced. The mechanical properties of the micro-resonator are strengthened by optimizing the parameters in these equations. The neutral plane tension, neutral plane bending, and electrostatic nonlinearity are considered in the mathematical model. First, the model is reduced and simplified by Galerkin discretization. Then the relationship between two section parameters under linear vibration is obtained by a Newton iteration. Finally, the vibration at small and large amplitudes were obtained using MMS and NMMS. The results were verified with the Runge-Kutta method, Simulink dynamic simulation and finite element method.

The key conclusions are as follows.

1. The potential energy and electrostatic force are greater when the microbeam intermediate thickness is thinner for a given applied voltage. Decreasing the thickness at the midpoint of the microbeam will increase the pull-in voltage and increase the likelihood of secondary pull-in. The pull-in position is biased toward the larger of λ_1 and $-\lambda_2$ when microbeam is asymmetric about the neutral plane, and the pull-in voltage is primarily determined by the larger of λ_1 and $-\lambda_2$.

2. When the micro-resonator vibrates linearly, the relationship between λ_1 and λ_2 is symmetric at $y = x$. The eigenfrequency and amplitude of the microbeam are studied when λ_1 and λ_2 correspond to linear vibration. Regarding the $y = x$ symmetric point in Figure 7, the microbeams have the same eigenfrequency and different amplitudes. The amplitude near the AC voltage source is larger.

3. The MMS should not be used as the amplitude increases. The frequency response obtained with MMS will exhibit weaker nonlinear softening. If the initial vibration exhibits hardening, the transition between hardening and softening will occur at a certain time as the vibration amplitude increases. The position of the transition point does not change as the amplitude continues to increase. At the transition point, the micro-resonator can jump to a higher stable branch and output more energy.

It is found that the static and dynamic behaviors of micro-resonator can be changed by adjusting the microbeam shape. Determining the appropriate geometry can improve the mechanical properties for given raw materials.

Author Contributions: J.F., C.L. and W.Z. conceived and designed the model; J.F., C.L. and J.H. contributed theoretical analysis; J.F., J.H. and S.H. analyzed the data; C.L., W.Z. and S.H. conducted the simulation; C.L. and J.F. wrote the paper.

Funding: The project supported by the National Natural Science Foundation of China (11602169, 11832002, 11702192, 11772218 and 11872044), Natural Science Foundation of Tianjin City (16JCQNJC04700 and 17JCYBJC18900) and Scientific Research Project of Tianjin Municipal Education Commission (2017KJ262).

Conflicts of Interest: The authors declare no conflict of interest.

Appendix A

$$p = \int_0^1 A(x)\phi'(x)^2 dx \tag{A1}$$

$$q = \int_0^1 A(x)\phi'(x)(y_1(x) + y_2(x))' dx \tag{A2}$$

$$A_{ij} = \int_0^1 (I(x)\phi''(x))''(1 - y_1(x))^i(1 + y_2(x))^j\phi(x)^{5-i-j} dx \tag{A3}$$

$$B_{ij} = \int_0^1 A(x)\phi(x)(1 - y_1(x))^i(1+y_2(x))^j\phi(x)^{5-i-j} dx \tag{A4}$$

$$C_{ij} = \int_0^1 \phi(x)(1 - y_1(x))^i(1 + y_2(x))^j\phi(x)^{5-i-j} dx \tag{A5}$$

$$D_{ij} = \int_0^1 \phi''(x)(1 - y_1(x))^i(1 + y_2(x))^j\phi(x)^{5-i-j} dx \tag{A6}$$

$$E_{ij} = \int_0^1 (y_1(x) + y_2(x))''(1 - y_1(x))^i(1+y_2(x))^j\phi(x)^{5-i-j} dx \tag{A7}$$

$$vdc_1 = \int_0^1 (1 + y_2(x))^2 - (1 - y_1(x))^2 dx \tag{A8}$$

$$vdc_2 = \int_0^1 2(1 + y_2(x) - y_1(x))\phi(x) dx \tag{A9}$$

$$vac_1 = \int_0^1 (1 + y_2(x))^2 dx \tag{A10}$$

$$vac_2 = \int_0^1 2(1 + y_2(x))\phi(x) dx \tag{A11}$$

$$vac_3 = \int_0^1 \phi(x)^2 dx \tag{A12}$$

$$g = B_{22} + 2(B_{21} - B_{12})u(t) + (B_{20} + B_{02} - 4B_{11})u(t)^2 + 2(B_{01} - B_{10})u(t)^3 + B_{00}u(t)^4 \tag{A13}$$

$$\mu = c'(C_{22} + 2(C_{21} - C_{12})u(t) + (C_{20} + C_{02} - 4C_{11})u(t)^2 + 2(C_{01} - C_{10})u(t)^3 + C_{00}u(t)^4) \tag{A14}$$

$$k_0 = -\frac{1}{2}NF_{22} - \alpha_1 v_{dc1} \tag{A15}$$

$$k_1 = A_{22} - N(D_{22} + E_{21} - E_{12}) - \frac{\alpha_2}{2}qE_{22} - \alpha_1 v_{dc2} \tag{A16}$$

$$k_2 = 2(A_{21} - A_{12}) - N(2D_{21} - 2D_{12} + \frac{1}{2}E_{20} + \frac{1}{2}E_{02} - 2E_{11}) - \alpha_2(\frac{1}{2}pE_{22} + qD_{22} + qE_{21} - qE_{12}) \tag{A17}$$

$$k_3 = \begin{aligned}[t] &A_{20} + A_{02} - 4A_{11} - N(D_{20} + D_{02} - 4D_{11} + E_{01} - E_{10}) - \alpha_2(pD_{22} + pE_{21} - pE_{12} + 2qD_{21} \\ &-2qD_{12} + \tfrac{1}{2}qE_{20} + \tfrac{1}{2}qE_{02} - 2qE_{11}) \end{aligned} \tag{A18}$$

$$k_4 = \begin{aligned}[t] &2A_{01} - 2A_{10} - N(2D_{01} - 2D_{10} + \tfrac{1}{2}E_{00}) - \alpha_2(2pD_{21} - 2pD_{12} + \tfrac{1}{2}pE_{20} + \tfrac{1}{2}pE_{02} \\ &-2pE_{11} + qD_{20} + qD_{02} - 4qD_{11} + qE_{01} - qE_{10}) \end{aligned} \tag{A19}$$

$$k_5 = A_{00} - ND_{00} - \alpha_2(pD_{20} + pD_{02} - 4pD_{11} + pE_{01} - pE_{10} + 2qD_{01} - 2qD_{10} + \frac{1}{2}qE_{00}) \tag{A20}$$

$$k_6 = -\alpha_2(2pD_{01} - 2pD_{10} + \frac{1}{2}pE_{00} + qD_{00}) \tag{A21}$$

$$k_7 = -\alpha_2 pD_{00} \tag{A22}$$

Appendix B

$$\chi = B_{22} + 2(B_{21} - B_{12})u_D + (B_{20} + B_{02} - 4B_{11})u_D^2 + 2(B_{01} - B_{10})u_D^3 + B_{00}u_D^4 \tag{A23}$$

$$a_m = (2(B_{21} - B_{12}) + 2(B_{20} + B_{02} - 4B_{11})u_D + 6(B_{01} - B_{10})u_D^2 + 4B_{00}u_D^3)/\chi \tag{A24}$$

$$a_n = (B_{20} + B_{02} - 4B_{11} + 6(B_{01} - B_{10})u_D + 6B_{00}u_D^2)/\chi \tag{A25}$$

$$\omega_n = ((k_1 + 2k_2 u_D + 3k_3 u_D^2 + 4k_4 u_D^3 + 5k_5 u_D^4 + 6k_6 u_D^5 + 7k_7 u_D^6)/\chi)^{0.5} \tag{A26}$$

$$a_q = (k_2 + 3k_3 u_D + 6k_4 u_D^2 + 10k_5 u_D^3 + 15k_6 u_D^4 + 21k_7 u_D^5)/\chi \tag{A27}$$

$$a_c = (k_3 u_D + 4k_4 u_D + 10k_5 u_D^2 + 20k_6 u_D^3 + 35k_7 u_D^4)/\chi \tag{A28}$$

$$f = 2\alpha_1\rho(v_{ac1} + v_{ac2}u_D + v_{ac2}u_D^2)/\chi \tag{A29}$$

References

1. Lai, Y.L.; Chang, L.H. Design of electrostatically actuated MEMS switches. *Colloids Surf. A* **2008**, *313*, 469–473. [CrossRef]
2. Tian, W.C.; Li, P.; Yuan, L.X. Research and Analysis of MEMS Switches in Different Frequency Bands. *Micromachines* **2018**, *9*, 185. [CrossRef] [PubMed]
3. Nayfeh, A.H.; Ouakad, H.M.; Najar, F.; Choura, S.; Abdel-Rahman, E.M. Nonlinear dynamics of a resonant gas sensor. *Nonlinear Dyn.* **2010**, *59*, 607–618. [CrossRef]
4. Ibrahim, A.; Younis, M.I. Simple fall criteria for MEMS sensors: Data analysis and sensor concept. *Sensors* **2014**, *14*, 12149–12173. [CrossRef]
5. Chu, H.M. Air damping models for micro- and nano-mechanical beam resonators in molecular-flow regime. *Vacuum* **2016**, *126*, 45–50. [CrossRef]
6. Chorsi, M.T.; Chorsi, H.T. Modeling and analysis of MEMS disk resonators. *Microsyst. Technol.* **2018**, *24*, 2517–2528. [CrossRef]
7. Chorsi, H.T.; Chorsi, M.T.; Gedney, S.D. A conceptual study of microelectromechanical disk resonators. *IEEE J. Multiscale Multiphys. Comput. Technol.* **2017**, *2*, 29–37. [CrossRef]
8. Xie, Y.; Li, S.S.; Lin, Y.W.; Ren, Z.; Nguyen, C.T.C. 1.52-GHz micromechanical extensional wine-glass mode ring resonators. *IEEE Trans. Ultrason. Ferroelectr. Freq. Control* **2008**, *55*, 890–907.
9. Chorsi, M.T.; Chorsi, H.T.; Gedney, S.D. Radial-contour mode microring resonators: Nonlinear dynamics. *Int. J. Mech. Sci.* **2017**, *130*, 258–266. [CrossRef]

10. Rezazadeh, G.; Madinei, H.; Shabani, R. Study of parametric oscillation of an electrostatically actuated microbeam using variational iteration method. *Appl. Math. Model.* **2012**, *36*, 430–443. [CrossRef]

11. Wang, L.; Hong, Y.Z.; Dai, H.L.; Ni, Q. Natural frequency and stability tuning of cantilevered CNTs conveying fluid in magnetic field. *Acta Mech. Solida Sin.* **2016**, *59*, 567–576. [CrossRef]

12. Deng, J.F.; Liu, C.; Hao, Y.P.; Liu, S.J.; Meng, F.J.; Xu, P.P. Control method of pull-in voltage on the MEMS inertial switch integrating actuator and sensor. *Microsyst. Technol.* **2017**, *23*, 4785–4795. [CrossRef]

13. Abdel-Rahman, E.M.; Younis, M.I.; Nayfeh, A.H. Characterization of the mechanical behavior of an electrically actuatedmicrobeam. *J. Micromech. Microeng.* **2002**, *12*, 759–766. [CrossRef]

14. Mobki, H.; Rezazadeh, G.; Sadeghi, M.; Vakili-Tahami, F.; Seyyed-Fakhrabadi, M. A comprehensive study of stability in an electro-statically actuated microbeam. *Int. J. Nonlinear Mech.* **2013**, *48*, 78–85. [CrossRef]

15. Li, L.; Zhang, Q.C.; Wang, W.; Han, J.X. Dynamic analysis and design of electrically actuated viscoelastic microbeams considering the scale effect. *Int. J. Nonlinear Mech.* **2017**, *90*, 21–31. [CrossRef]

16. Han, J.X.; Zhang, Q.C.; Wang, W. Static bifurcation and primary resonance analysis of a MEMS resonator actuated by two symmetrical electrodes. *Nonlinear Dyn.* **2015**, *80*, 1585–1599. [CrossRef]

17. Li, L.; Zhang, Q.C.; Wang, W.; Han, J.X. Nonlinear coupled vibration of electrostatically actuated clamped–clamped microbeams under higher-order modes excitation. *Nonlinear Dyn.* **2017**, *90*, 1593–1606. [CrossRef]

18. Hu, K.M.; Zhang, W.M.; Dong, X.J.; Peng, Z.K.; Meng, G. Scale effect on tension-induced intermodal coupling in nanome-chanical resonators. *J. Vib. Acoust.* **2015**, *137*, 021008. [CrossRef]

19. Joglekar, M.M.; Pawaskar, D.N. Shape optimization of electrostatically actuated microbeams for extending static and dynamic operating ranges. *Struct. Multidiscip. Optim.* **2012**, *46*, 871–890. [CrossRef]

20. Trivedi, R.R.; Joglekar, M.M.; Shimpi, R.R.; Pawaskar, D.N. Shape optimization of electrostatically actuated micro cantilever beam with extended travel range using simulated annealing. In Proceedings of the World Congress on Engineering, London, UK, 6–8 July 2011; pp. 2042–2047.

21. Zhang, S.; Zhang, W.M.; Peng, Z.K.; Meng, G. Dynamic characteristics of electrostatically actuated shape optimized variable geometry microbeam. *Shock Vib.* **2015**, *2015*, 867171. [CrossRef]

22. Kuang, J.H.; Chen, C.J. Dynamic characteristics of shaped microactuators solved using the differential quadrature method. *J. Micromech. Microeng.* **2004**, *14*, 647–655. [CrossRef]

23. Alsaleem, F.; Younis, M.I. Integrity analysis of electrically actuated resonators with delayed feedback controller. *J. Dyn. Syst. Meas. Control* **2011**, *133*, 031011. [CrossRef]

24. Zhang, M.W.; Meng, G. Nonlinear dynamic analysis of electrostatically actuated resonant MEMS sensors under parametric excitation. *IEEE Sens. J.* **2007**, *7*, 370–380. [CrossRef]

25. Morrison, T.M.; Rand, R.H. 2:1 Resonance in the delayed nonlinear Mathieu equation. *Nonlinear Dyn.* **2007**, *50*, 341–352. [CrossRef]

26. Masri, K.M.; Shao, S.; Younis, M.I. Delayed feedback controller for microelectromechanical systems resonators undergoing large motion. *J. Vib. Control* **2015**, *21*, 2604–2615. [CrossRef]

27. Najar, F.; Choura, S.; El-Borgi, S.; Abdel-Rahman, E.M.; Nayfeh, A.H. Modeling and design of variable-geometry electrostatic microactuators. *J. Micromech. Microeng.* **2005**, *15*, 419–429. [CrossRef]

28. Chen, T. *The Research of Nonlinear Dynamic Characteristics of Electrostatically Actuated Micro-Bridges*; Tianjin University: Tianjin, China, 2016.

29. Feng, J.J.; Liu, C.; Zhang, W.; Hao, S.Y. Static and Dynamic Mechanical Behaviors of Electrostatic MEMS Resonator with Surface Processing Error. *Micromachines* **2018**, *9*, 34. [CrossRef]

30. Younis, M.I.; Abdel-Rahman, E.M.; Nayfeh, A.A. Reduced-Order Model for Electrically Actuated Microbeam-Based MEMS. *J. Microelectromech. Syst.* **2003**, *12*, 672–680. [CrossRef]

31. Shao, S.; Masri, K.M.; Younis, M.I. The effect of time-delayed feedback controller on an electrically actuated resonator. *Nonlinear Dyn.* **2013**, *74*, 257–270. [CrossRef]

32. Han, J.X.; Qi, H.J.; Jin, G. Mechanical behaviors of electrostatic microresonators with initial offset imperfection: Qualitative analysis via time-varying capacitors. *Nonlinear Dyn.* **2018**, *91*, 269–295. [CrossRef]

33. Han, J.X.; Jin, G.; Zhang, Q.C.; Wang, W.; Li, B.Z.; Qi, H.J.; Feng, J.J. Dynamic evolution of a primary resonance MEMS resonator under prebuckling pattern. *Nonlinear Dyn.* **2018**, *93*, 2357–2378. [CrossRef]

Article

Temperature-Insensitive Structure Design of Micromachined Resonant Accelerometers

Yonggang Yin [iD], **Zhengxiang Fang, Yunfeng Liu * and Fengtian Han ***[iD]

Department of Precision Instrument, Tsinghua University, Beijing 100084, China;
yinyg14@mails.tsinghua.edu.cn (Y.Y.); fang-zx16@mails.tsinghua.edu.cn (Z.F.)
* Correspondence: yfliu@mail.tsinghua.edu.cn (Y.L.); hft@mail.tsinghua.edu.cn (F.H.);
 Tel.: +86-10-62792119 (Y.L.); +86-10-62798645 (F.H.)

Received: 27 February 2019; Accepted: 26 March 2019; Published: 30 March 2019

Abstract: Micromachined resonant accelerometers (MRAs), especially those devices fabricated by silicon on glass technology, suffer from temperature drift error caused by inherent thermal stress. This paper proposes two structure designs to attenuate the effect of thermal stress. The first MRA structure is realized by optimizing the locations of the bonding anchors and utilizing a special-shaped substrate to isolate the thermal stress generated during the die attach process. The second structure is designed using an isolation frame fixed by a single anchor to replace all dispersed anchors associated with the suspension beams and micro-levers. Simulated and experimental results show that both of the MRA structures can effectively reduce the thermal stress effect. The experimental results on one MRA prototype indicate that the differential temperature sensitivity reduces down to 1.9 μg/°C and its 15-day bias stability reaches 1.4 μg.

Keywords: microelectromechanical systems (MEMS); resonant accelerometer; silicon on glass (SOG); temperature sensitivity; thermal stress

1. Introduction

With the rapid development of microelectromechanical systems (MEMS) technology over three decades, micromachined accelerometers have shown potential for high-precision applications, such as strategic-grade navigation [1], seismology [2–5] and gravimetry [6,7]. Based on the force-frequency characteristics of high-Q resonators, micromachined resonant accelerometer (MRA) is very attractive due to its high resolution and large dynamic range [8–10]. However, the temperature-dependent drifts of the bias and scale factor limit the long-term stability of those high-performance MRAs. There are two major sources of temperature drift dominated by the sensing structure [11]. Firstly, the Young's modulus of silicon [12] varies with temperature change, which will directly result in the natural frequency drift of the resonator [13]. Secondly, the mismatch of the coefficients of thermal expansion (CTEs) between the silicon structure and its substrate will inevitably generate thermal stress.

Several methods have been reported in the literature to reduce the two errors mentioned above. For silicon micromechanical resonators used as frequency and timing devices, the temperature drift of Young's modulus can be compensated effectively by doping the silicon with dopants [14,15] or using a composite temperature-insensitive structure, such as silicon dioxide and silicon, whose Young's modulus changes with temperature in opposite directions [13,16]. Nevertheless, the methods above are not generally utilized in MRAs because the temperature drift of Young's modulus can be greatly decreased by a differential structure design comprising two symmetric resonators [17,18].

Currently, the effect of the thermal stress becomes the main limitation on the MRA's output stability. Michel [19] designed a silicon resonator on a glass substrate and proposed a method to counteract the effect of the thermal stress and the Young's modulus by adjusting the distance between resonator anchors. Since the change in Young's modulus will make the resonant frequency decrease as the temperature rises, the frequency drift induced by thermal stress should be positive by design. In a similar manner, Myers et al. [20] designed a silicon carbide resonator on a single-crystalline silicon substrate to counteract the two kinds of drifts. However, the die attach process during the device packaging will generate unpredictable thermal stress [21], which could deviate the theoretical result of the temperature insensitive point greatly. Zhao et al. [22] and Zhang et al. [23] proposed a frame structure to decrease the amount of anchors. In these MRAs, the resonators and the micro-levers are connected to the fixed frame instead of anchors so that the thermal stress from the substrate will not act on the resonators directly. However, the two resonators in this design still suffer from large thermal stress because the deformation of the frame is very sensitive to the temperature change. Yang et al. [24] proposed an isolation platform for inertial sensors, which can reduce the thermal stress from the package and printed circuit board. However, the adhesive layer over the platform still has a remarkable effect on the sensor structure. Shin et al. [25–27] demonstrated a differential resonant accelerometer with a single-point anchor. This design can greatly eliminate frequency fluctuations, due to package stress. Nevertheless, the structure may be fragile and suffers from large cross-axis coupling error as a small anchor was used to support the proof mass where the resonant beam acts as a cantilever.

Our preliminary results on the temperature-insensitive structure with thermal stress isolation were presented in Reference [28]. In this paper, more theoretical analysis and the latest experimental results are introduced and demonstrated the low temperature drift performance. Moreover, a novel structure is also proposed using an isolation frame fixed by a single anchor to replace all dispersed anchors associated with the suspension beams and micro-levers. These two structure designs aim to minimize the effect of thermal stress and improve the temperature drift consistency of the two resonators operating in a differential output.

2. Operation Principle of the MRA

The structure of our MRA is fully symmetric and half of the schematic structure is illustrated in Figure 1. The movable silicon structure is bonded to a glass substrate via a set of anchors. The input acceleration along the sensitive axis (*x*-axis) will make the proof mass displace away from its nominal position. Then the inertial force is amplified by the micro-lever system in order to achieve high sensitivity. Finally, the double-ended tuning fork (DETF) resonator in Figure 1 is stretched, which will increase its natural frequency. On the other hand, the symmetric DETF is compressed and its natural frequency will be decreased. Within a certain measurement range, the differential frequency of the two resonators is proportional to the input acceleration. This differential output will cancel most of the common error sources of the two resonators. Note that the two resonant beams in each DETF vibrate in reversed-phase mode, which can improve the Q-factor of the resonators.

Consider the resonant beam as a fixed-fixed beam, its resonant frequency can be given by

$$f_0 = \frac{1}{2\pi}\sqrt{\frac{199EI}{ml^3}}, \tag{1}$$

where E is the Young's modulus of silicon, I is the inertia moment of the cross section, l is the length of the beam and m is the equivalent mass. Then the temperature coefficient of the resonant frequency caused by the Young's modulus is

$$C_E = \frac{df_0}{dT} = \frac{f_0}{2}\frac{dE}{EdT}, \tag{2}$$

where T is the temperature. When an external force P is applied along the beam, the resonant frequency will be

$$f_P = f_0\sqrt{1 + \frac{Pl^2}{4\pi^2EI}} \approx f_0(1 + \frac{Pl^2}{8\pi^2EI}). \tag{3}$$

The temperature coefficient of frequency change caused by the external force is

$$C_P = \frac{df_P}{dT} = \frac{f_0 l^2}{8\pi^2 EI}\frac{dP}{dT}.$$

(4)

In this work, the MRA devices were fabricated based on silicon on glass (SOG) process [1,29]. Since the CTE of the glass substrate (α_g) is slightly larger than that of the silicon structure (α_s), the x-axis distance between the anchors (δ) will increase as the temperature rises. In this case, the tensile thermal force will be generated in the resonant beam, as modeled in

$$P = k\delta(\alpha_g - \alpha_s)T.$$

(5)

where k is the axial stiffness of the resonant beam.

Worse still, during the subsequent die attach process by packaging the glass substrate on a ceramic base, large thermal expansion of the adhesive layer will deteriorate the theoretical frequency drift results greatly. Its effect can be equivalent to increase the CTE of the glass substrate (α_g) in (5). Additionally, it is difficult to maintain a uniform thickness of the adhesive layer, which will increase the differential frequency error, due to inconsistent temperature responses of the two resonators.

In order to reduce the effect of the thermal stress and improve the temperature drift consistency of the two differential resonators in the MRA device, two structure designs are proposed in this work.

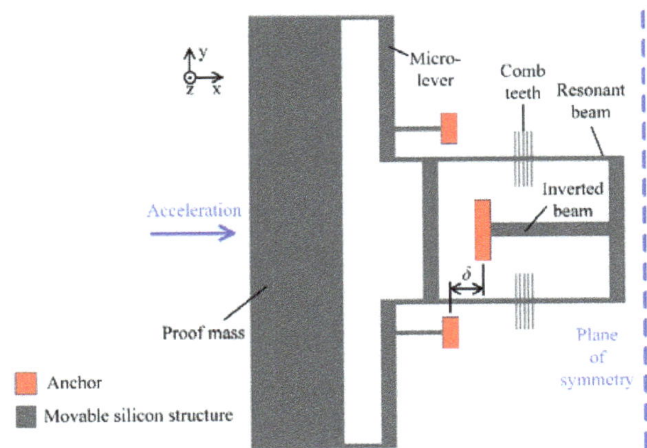

Figure 1. A schematic of the sensing structure (half of the symmetric device).

3. Design A: Anchors' Location Optimization and Isolation Substrate

In the first design, the locations of the bonding anchors are optimized to decrease the thermal stress between the glass substrate and the silicon structure. In principle, the thermal stress coefficient (k) is dominated by δ. If $\delta = 0$, which can be implemented by using an inverted beam design (as shown in Figure 1), the thermal stress along the acceleration-sensitive axis can be decreased greatly. On the other hand, the layout of the suspension beams also influences the thermal stress. The four anchors of the suspension beams should be arranged symmetrically, relative to the expansion center, and be close to each other. Improved structure design is illustrated schematically in Figure 2, where the condition $\delta = 0$ holds.

In addition, a dedicated H-shaped isolation structure is proposed to reduce the thermal stress produced by the adhesive layer during the die attach process, as shown in Figure 3. Compared with a traditional rectangular glass substrate, the H-shaped glass substrate has a narrow isolation beam. One side of the H-shaped substrate is attached to the ceramic base via adhesive. The other side, which provides the bonding area of the silicon structure, is separated from the adhesive layer by the isolation beam. In this way, the silicon structure and the glass substrate below can virtually expand or shrink freely without the constraint of the ceramic base.

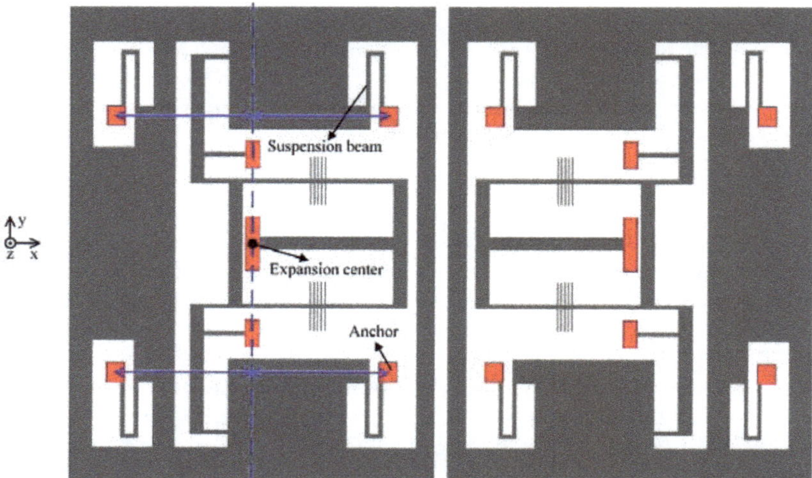

Figure 2. The schematic structure with optimized anchor locations.

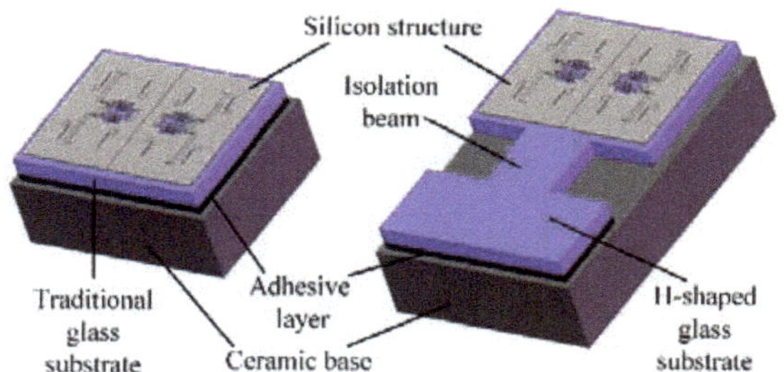

Figure 3. Comparison of micromachined resonant accelerometers (MRAs) with the traditional substrate and H-shaped isolation substrate.

The temperature characteristics of design A are simulated by finite element analysis. To focus on the effect of the thermal stress, the Young's modulus of all the materials are set to be constant while the temperature changes. The temperature drifts of the resonant frequency are respectively simulated before and after the glass substrate is fixed on the ceramic base by an adhesive. The simulation results show that the theoretical temperature drift caused by the thermal expansion of silicon and glass is only −0.0136 Hz/°C before die attach, as shown in Figure 4. After die attach, the frequency drift will increase up to −0.1860 Hz/°C if we use a traditional rectangular glass substrate. However, the frequency drift coefficient will change a little bit more, only −0.0393 Hz/°C if the H-shaped substrate is applied. The simulated thermal stress distribution of design A is shown in Figure 5. The finite element simulation results indicate that the H-shaped substrate can effectively isolate the thermal stress generated by the adhesive layer and the ceramic base. The thermal stress attenuates rapidly in the narrow isolation beam so that the residual stress in the silicon structure area can maintain a low level.

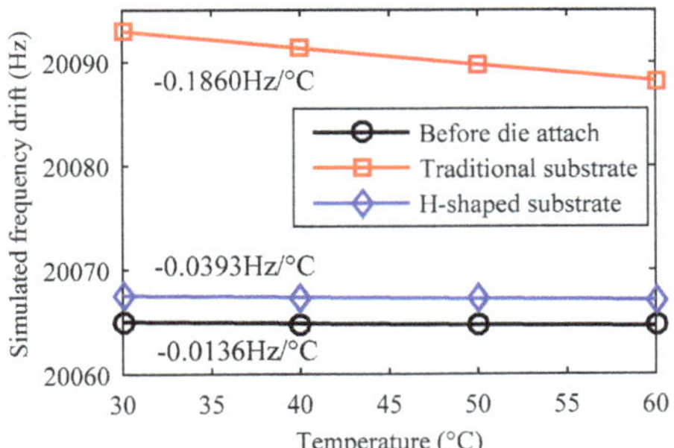

Figure 4. The simulated temperature drift of a single resonator's natural frequency dominated by the thermal stress in design A.

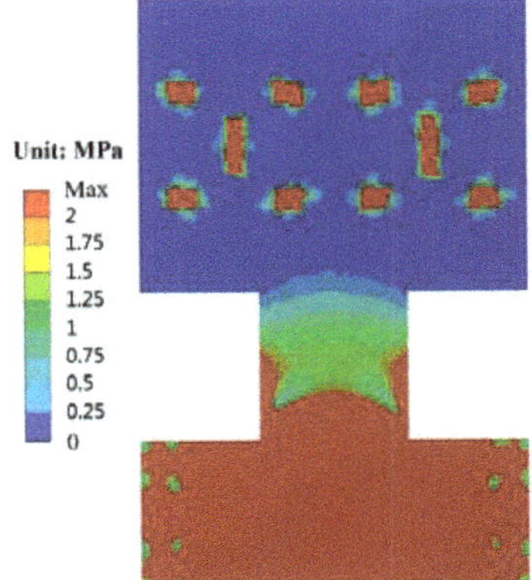

Figure 5. The simulated thermal stress distribution of the H-shaped substrate after die attach.

4. Design B: Single Anchored Isolation Frame

Although optimizing the location of anchors can remove most of the thermal stress along the sensitive *x*-axis, the stress along the *y*-axis still exists. Moreover, the H-shaped glass substrate can effectively decrease the package stress at the expense of extra processing steps, such as laser cutting and chip cleaning, which will also cause yield loss.

Alternatively, a novel structure is further proposed to reduce the thermal stress and avoid the disadvantages of design A. In design B, all the traditional anchors of the suspension beams and micro-levers are removed and replaced by a rectangular isolation frame, as illustrated in Figure 6. Note that the structure is symmetric and the whole movable silicon structure is bonded to the substrate through a single anchor located in the center. It is noted that the suspension beams are directly connected to the isolation frame. The fulcrum of the micro-lever is fixed to the frame by a relatively wide and long beam. The inverted beam is no longer necessary and the resonator is connected to the other side of the frame away from the anchor.

Since the single anchor will not generate relative location change as above multiple anchors may occur, in principle the structure in design B can attenuate the thermal stress both from the substrate and the die attach process. Finite element simulation shows that the theoretical temperature drift caused by the thermal expansion of silicon and glass is only -0.0003 Hz/°C, as shown in Figure 7. This result is much smaller than that in design A (-0.0136 Hz/°C). After die attach, the frequency drift will be only -0.0018 Hz/°C, even using a traditional rectangular substrate. The simulated thermal stress distribution of design B is shown in Figure 8. Considering the desirable bonding strength, the anchor area should not be too small, e.g., 1.72 mm^2 in this work. The thermal stress is mainly generated inside the anchor area. Since the sensitive resonator and the suspension beams are configured to be far away from the anchor, the thermal stress in the resonant beam is negligible.

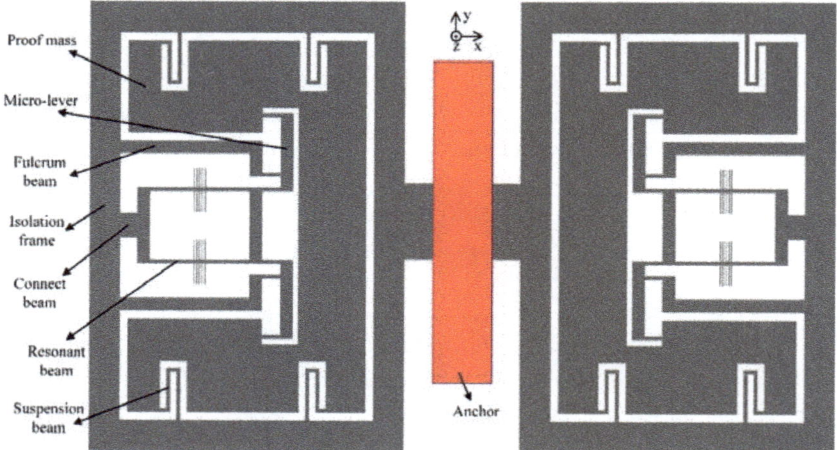

Figure 6. The schematic structure with single-anchored isolation frame.

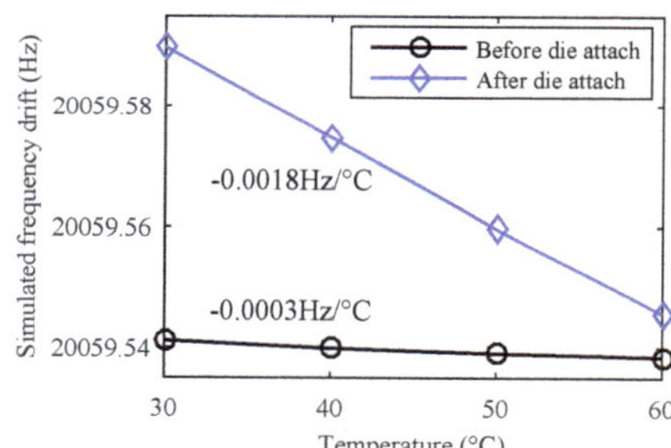

Figure 7. The simulated temperature drift of a single resonator's natural frequency dominated by the thermal stress in design B.

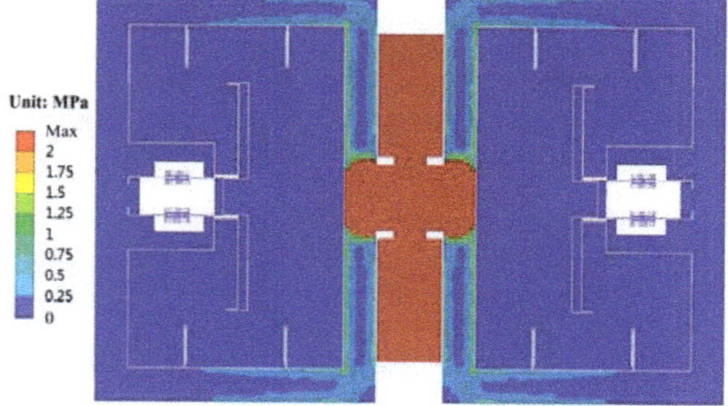

Figure 8. The simulated thermal stress distribution of the silicon structure after die attach in design B.

In designs A and B, the frequency drift caused by the Young's modulus itself is simulated to be -0.5598 Hz/°C. This will dominate the total temperature drift of each single resonator. Benefiting from the full symmetric design of two resonators, the temperature drift of the differential frequency output will be reduced greatly.

Besides, the simulated modal frequencies of the H-shaped substrate and the single anchor frame are higher than 7.2 kHz and 1.7 kHz respectively, which are much larger than the operating bandwidth of the MRA (typically less than 0.5 kHz). Therefore, the MRA measurement is virtually undisturbed by these vibration modes of the two isolation structures.

5. Fabrication and Experiments

The MRA devices designed by both methods A and B were fabricated and tested experimentally. The silicon-glass structure was sealed in a hermetic metallic package and maintained at a high vacuum by a getter inside the package. The photographs of the SOG devices and associated package are shown in Figure 9. The degree of vacuum maintained inside the package is estimated at 0.1 Pa and the measured Q-factor of the resonator is up to 2.5×10^5. The device is wire-bonded through the bond pads distributed on the edges of the glass substrate.

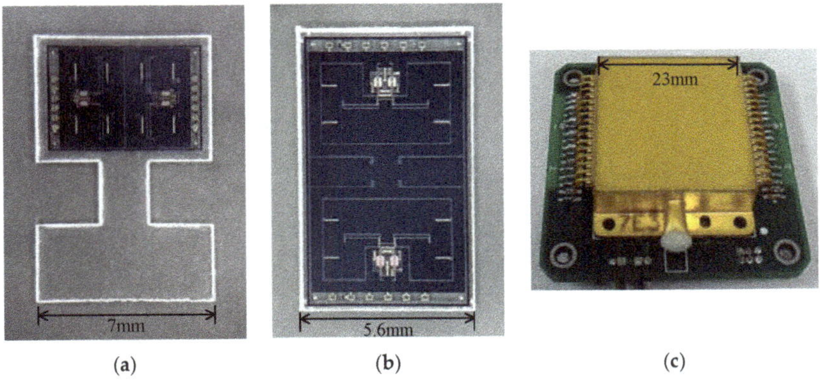

Figure 9. Photographs of the fabricated devices: (**a**) the structure in design A, (**b**) the structure in design B, (**c**) the vacuum packaged MRA device on the interface circuit board.

The block diagram of the interface electronics is shown in Figure 10. The vibration of the resonator will cause the change of the capacitance between the comb teeth. Then the differential capacitance is detected by a ring-type diode based demodulation circuit [30]. The vibration amplitude is stabilized by automatic amplitude control (AAC). With a 90° phase shifter, the self-oscillation loop is established so the resonator will be locked at its resonant frequency. The frequency signal is captured by a commercial frequency counter (53230A, Keysight Technologies, Santa Rosa, CA, USA) and recorded by a computer in the following tests.

The measured temperature drift of three types of MRA devices (A1~A2, A3~A4 and B1~B2) are shown in Table 1. The devices A1~A4 are designed by method A and the devices B1~B2 are designed by method B. Note that the scale factors (SFs) of these devices vary widely because the force amplification factors of the micro-lever system are different in design. Imperfect fabrication contributes to the SF discrepancy in the same design of devices. Considering the full scale range of these accelerometers are designed to be greater than ±15 g, the scale factors listed in Table 1 were measured by setting the input axis of the accelerometer to different gravity vectors of +1 g and −1 g, respectively. In addition, f_1 and f_2 are the resonant frequency of the two resonators in one MRA device, respectively. Let TCF1 and TCF2 be the temperature coefficients of the frequency of the two resonators, the differential temperature coefficient of acceleration output is calculated by TCA = (TCF1 − TCF2)/SF. The measured TCF of a single resonator ranges from −0.5924 Hz/°C to −0.5330 Hz/°C, which agrees well with the simulated predication (−0.5598 Hz/°C). The sizes of all the resonators were designed to be the same while imperfect fabrication in different batches could cause the discrepancy. Considering the SF results, all the differential TCAs are lower than 13.9 µg/°C and the best TCA reaches as low as 1.9 µg/°C. The temperature drifts of the prototypes A1 and B1 are shown in Figure 11. The experimental results confirm the two proposed structures are effective to reduce thermal stress. The residual temperature drift is mainly caused by the fabrication discrepancy of the two resonators.

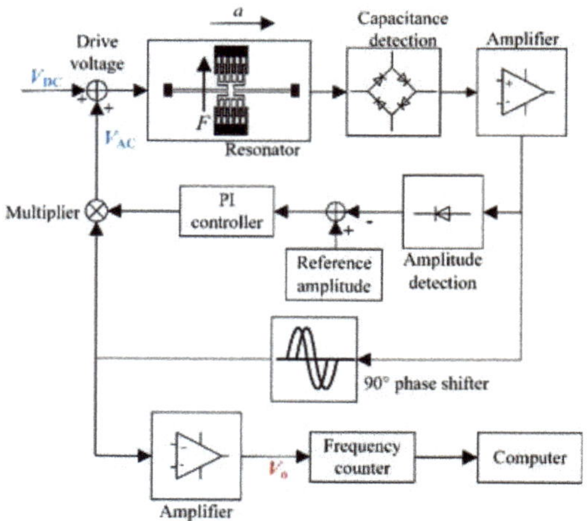

Figure 10. The block diagram of the MRA interface electronics.

Table 1. Temperature drift of three types of MRA devices.

Serial Number	f_1 (Hz)	f_2 (Hz)	TCF1 (Hz/°C)	TCF2 (Hz/°C)	SF (Hz/g)	TCA (µg/°C)
A1	19,495	19,450	−0.5566	−0.5573	364	−1.9
A2	19,516	19,524	−0.5563	−0.5547	361	−4.4
A3	20,527	20,517	−0.5907	−0.5873	244	−13.9
A4	20,528	20,584	−0.5897	−0.5924	256	10.5
B1	20,704	20,667	−0.5339	−0.5330	203	4.4
B2	21,008	20,976	−0.5369	−0.5355	196	7.1

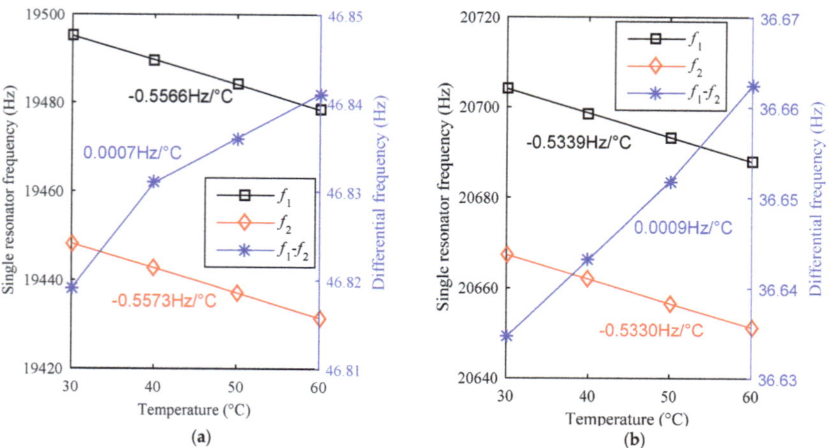

Figure 11. The measured temperature drifts of two MRA prototypes: (**a**) prototype A1 designed by method A, (**b**) prototype B1 designed by method B.

Long-term bias stability is one of the most crucial performance requirements for high-precision accelerometers. We have recorded the bias stability of the prototype A1 for 15 days, as shown in Figure 12. The device is mounted at 0 g position and the test environment is under room temperature. The frequency output is acquired at a sampling rate of 1 Hz, where no temperature compensation or data smoothing is used in the plotted curves. After the MRA reached thermal equilibrium within two hours, the continuous measurement lasted 15 days. The standard deviation of the 15-day data is 1.4 µg, which is the lowest bias stability ever reported for MEMS accelerometers without temperature compensation or temperature control. The measured results indicate that the MRAs presented in this work exhibit expected temperature-insensitive characteristics and excellent long-term stability.

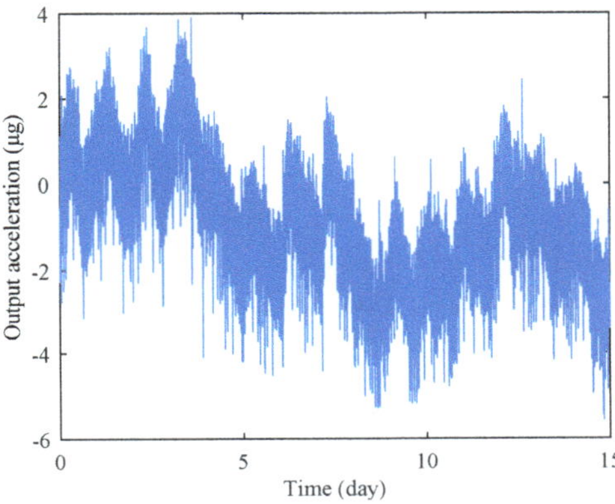

Figure 12. Long-time bias stability measurement of prototype A1 under room temperature.

The Allan deviation is a generally used method to evaluate the noise and drift error of various accelerometers. The minimum of the Allan deviation indicates the bias instability which limits the attainable resolution of the device. Along with the low noise design [31], the MRA shows high overall performance in Allan deviation measurement, as shown in Figure 13. The sampling frequency is set at 1 Hz and the total number of recorded data points is 60,000. The result shows a bias instability of 76 ng at 27 s averaging time, which is the best bias instability among reported MEMS resonant accelerometers.

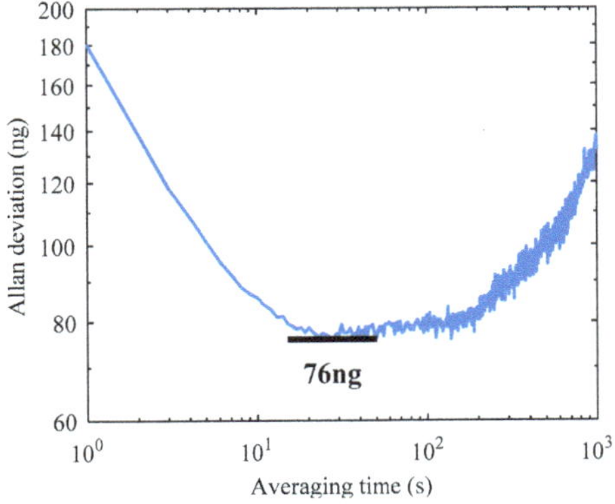

Figure 13. Allan deviation measurement of prototype A1.

6. Conclusions and Discussion

This paper presents two MRA structure designs to reduce the thermal stress effect. One design optimizes the anchor locations and utilizes an H-shaped isolation glass substrate instead of a classical rectangular substrate. The second design utilizes a single-anchored isolation frame to replace all the anchors connected to the suspension beams and micro-levers. Experimental results show that both of the two methods can effectively reduce the thermal stress generated in the microfabrication and device packaging process. The temperature drift coefficient of acceleration measurements is reduced down to 1.9 µg/°C. The bias stability is greatly improved where the 15-day result is 1.4 µg and the bias instability reaches 76 ng. The proposed structures offer a promising solution for the design of temperature-insensitive MRA devices. The experimental results are attractive for high-precision navigation and seismic measurement applications. In order to evaluate overall performance potential, future work will attempt to measure the Earth tides using our MRA prototypes.

Author Contributions: Conceptualization, Y.Y. and F.H.; methodology, Y.Y.; software, Z.F.; validation, Y.L.; formal analysis, Y.L.; investigation, Y.Y.; resources, F.H.; data curation, Y.Y.; writing—original draft preparation, Y.Y.; writing—review and editing, F.H., Z.F. and Y.L.; visualization, Y.L.; supervision, F.H.; project administration, F.H.; funding acquisition, F.H.

Funding: The work was funded by National Natural Science Foundation of China, Grant No. 41774189.

References

1. Hopkins, R.; Miola, J.; Setterlund, R.; Dow, B.; Sawyer, W. The silicon oscillating accelerometer: A high-performance MEMS accelerometer for precision navigation and strategic guidance applications. In Proceedings of the 61st Annual Meeting of the Institute of Navigation, Cambridge, MA, USA, 27–29 June 2005; pp. 1043–1052.

2. Zou, X.; Thiruvenkatanathan, P.; Seshia, A.A. A seismic-grade resonant MEMS accelerometer. *J. Microelectromech. Syst.* **2014**, *23*, 768–770.

3. Pike, W.; Delahunty, A.; Mukherjee, A.; Dou, G.; Liu, H.; Calcutt, S.; Standley, I. A self-levelling nano-g silicon seismometer. In Proceedings of the SENSORS, Valencia, Spain, 2–5 November 2014; pp. 1599–1602.

4. Zou, X.; Seshia, A.A. A high-resolution resonant MEMS accelerometer. In Proceedings of the 2015 Transducers-2015 18th International Conference on Solid-State Sensors, Actuators and Microsystems (TRANSDUCERS), Anchorage, AK, USA, 21–25 June 2015; pp. 1247–1250.

5. Wang, S.; Wei, X.; Zhao, Y.; Jiang, Z.; Shen, Y. A MEMS resonant accelerometer for low-frequency vibration detection. *Sens. Actuators A Phys.* **2018**, *283*, 151–158. [CrossRef]

6. Middlemiss, R.; Samarelli, A.; Paul, D.; Hough, J.; Rowan, S.; Hammond, G. Measurement of the Earth tides with a MEMS gravimeter. *Nature* **2016**, *531*, 614. [CrossRef] [PubMed]

7. Middlemiss, R.P. *A Practical MEMS Gravimeter*; University of Glasgow: Scotland, UK, 2016.

8. Su, S.X.; Yang, H.S.; Agogino, A.M. A resonant accelerometer with two-stage microleverage mechanisms fabricated by SOI-MEMS technology. *IEEE Sens. J.* **2005**, *5*, 1214–1223. [CrossRef]

9. He, L.; Xu, Y.P.; Palaniapan, M. A CMOS Readout Circuit for SOI Resonant Accelerometer With 4-µg Bias Stability and 20-µg/$\sqrt{Hz}$ Resolution. *IEEE J. Solid-State Circuits* **2008**, *43*, 1480–1490. [CrossRef]

10. Tocchio, A.; Caspani, A.; Langfelder, G. Mechanical and electronic amplitude-limiting techniques in a MEMS resonant accelerometer. *IEEE Sens. J.* **2012**, *12*, 1719–1725. [CrossRef]

11. Dong, J.-H.; Qiu, A.-P.; Shi, R. Temperature influence mechanism of micromechanical silicon oscillating accelerometer. In Proceedings of the 2011 IEEE Power Engineering and Automation Conference, PEAM 2011, Wuhan, China, 8–9 September 2011; pp. 385–389.

12. Hopcroft, M.A.; Nix, W.D.; Kenny, T.W. What is the Young's Modulus of Silicon? *J. Microelectromech. Syst.* **2010**, *19*, 229–238. [CrossRef]

13. Melamud, R.; Chandorkar, S.A.; Kim, B.; Lee, H.K.; Salvia, J.C.; Bahl, G.; Hopcroft, M.A.; Kenny, T.W. Temperature-insensitive composite micromechanical resonators. *J. Microelectromech. Syst.* **2009**, *18*, 1409–1419. [CrossRef]

14. Samarao, A.K.; Ayazi, F. Temperature compensation of silicon resonators via degenerate doping. *IEEE Trans. Electron Devices* **2012**, *59*, 87–93. [CrossRef]

15. Ng, E.J.; Hong, V.A.; Yang, Y.; Ahn, C.H.; Everhart, C.L.; Kenny, T.W. Temperature dependence of the elastic constants of doped silicon. *J. Microelectromech. Syst.* **2015**, *24*, 730–741. [CrossRef]

16. Tabrizian, R.; Casinovi, G.; Ayazi, F. Temperature-Stable Silicon Oxide (SilOx) Micromechanical Resonators. *IEEE Trans. Electron. Devices* **2013**, *60*, 2656–2663. [CrossRef]

17. Comi, C.; Corigliano, A.; Langfelder, G.; Longoni, A.; Tocchio, A.; Simoni, B. A resonant microaccelerometer with high sensitivity operating in an oscillating circuit. *J. Microelectromech. Syst.* **2010**, *19*, 1140–1152. [CrossRef]

18. Park, U.; Rhim, J.; Jeon, J.U.; Kim, J. A micromachined differential resonant accelerometer based on robust structural design. *Microelectron. Eng.* **2014**, *129*, 5–11. [CrossRef]

19. St Michel, N.A. *Force Multiplier in a Microelectromechanical Silicon Oscillating Accelerometer*; Massachusetts Institute of Technology: Cambridge, MA, USA, 2000.

20. Myers, D.R.; Azevedo, R.G.; Chen, L.; Mehregany, M.; Pisano, A.P. Passive substrate temperature compensation of doubly anchored double-ended tuning forks. *J. Microelectromech. Syst.* **2012**, *21*, 1321–1328. [CrossRef]

21. Zhang, X.; Park, S.; Judy, M.W. Accurate assessment of packaging stress effects on MEMS sensors by measurement and sensor–package interaction simulations. *J. Microelectromech. Syst.* **2007**, *16*, 639–649. [CrossRef]

22. Zhao, Y.; Zhao, J.; Wang, X.; Xia, G.M.; Qiu, A.P.; Su, Y.; Xu, Y.P. A Sub-µg Bias-Instability MEMS Oscillating Accelerometer with an Ultra-Low-Noise Read-Out Circuit in CMOS. *IEEE J. Solid-State Circuits* **2015**, *50*, 2113–2126. [CrossRef]

23. Zhang, J.; Wang, Y.; Zega, V.; Su, Y.; Corigliano, A. Nonlinear dynamics under varying temperature conditions of the resonating beams of a differential resonant accelerometer. *J. Micromech. Microeng.* **2018**, *28*, 075004. [CrossRef]

24. Yang, D.M.; Najafi, K.; Lemmerhirt, D.F.; Mitchell, J. A micro thermal and stress isolation platform for inertial sensors. In Proceedings of the 2018 IEEE International Symposium on Inertial Sensors and Systems (INERTIAL), Moltrasio, Italy, 26–29 March 2018; pp. 1–4.

25. Shin, D.D.; Ahn, C.H.; Chen, Y.; Christensen, D.L.; Flader, I.B.; Kenny, T.W. Environmentally robust differential resonant accelerometer in a wafer-scale encapsulation process. In Proceedings of the 2017 IEEE 30th International Conference on Micro Electro Mechanical Systems (MEMS), Las Vegas, NV, USA, 22–26 January 2017; pp. 17–20.

26. Shin, D.D.; Chen, Y.; Flader, I.B.; Kenny, T.W. Epitaxially encapsulated resonant accelerometer with an on-chip micro-oven. In Proceedings of the 2017 19th International Conference on Solid-State Sensors, Actuators and Microsystems (TRANSDUCERS), Kaohsiung, Taiwan, 18–22 June 2017; pp. 595–598.

27. Shin, D.D.; Chen, Y.; Flader, I.B.; Kenny, T.W. Temperature compensation of resonant accelerometer via nonlinear operation. In Proceedings of the 2018 IEEE Micro Electro Mechanical Systems (MEMS), Belfast, UK, 21–25 January 2018; pp. 1012–1015.

28. Yin, Y.; Fang, Z.; Dong, J.; Liu, Y.; Han, F. A Temperature-Insensitive Micromachined Resonant Accelerometer with Thermal Stress Isolation. In Proceedings of the 2018 IEEE SENSORS, New Delhi, India, 28–31 October 2018; pp. 1–4.

29. Huang, L.; Yang, H.; Gao, Y.; Zhao, L.; Liang, J. Design and implementation of a micromechanical silicon resonant accelerometer. *Sensors* **2013**, *13*, 15785–15804. [CrossRef] [PubMed]

30. Yan, B.; Liu, Y.; Dong, J. The optimization of drive and sense circuit in silicon micro-machined resonant accelerometer. In Proceedings of the 2016 IEEE Chinese Guidance, Navigation and Control Conference (CGNCC), Nanjing, China, 12–14 August 2016; pp. 2058–2063.

31. Yin, Y.G.; Fang, Z.X.; Han, F.T.; Yan, B.; Dong, J.X.; Wu, Q.P. Design and test of a micromachined resonant accelerometer with high scale factor and low noise. *Sens. Actuator A Phys.* **2017**, *268*, 52–60. [CrossRef]

 *sensors*

Article

Design of the Squared Daisy: A Multi-Mode Energy Harvester, with Reduced Variability and a Non-Linear Frequency Response

Mathieu Gratuze, Abdul Hafiz Alameh and Frederic Nabki *

Department of Electrical Engineering, École de Technologie Supérieure, Montréal, QC H3C 1K3, Canada
* Correspondence: frederic.nabki@etsmtl.ca

Received: 1 July 2019; Accepted: 21 July 2019; Published: 24 July 2019

Abstract: With the rise of the Internet of Things (IoT) and the ever-increasing number of integrated sensors, the question of powering these devices represents an additional challenge. The traditional approach is to use a battery; however, harvesting energy from the environment seems to be the most practical approach. To that end, the use of piezoelectric MEMS energy has been proven as a potential power source in a wide range of applications. In this work, a proof of concept for a new architecture for MEMS energy harvesters is presented. The influence of the dimensions and different characteristics of these designs is discussed. These designs have been proven to be resilient to process variation thanks to their unique architecture. This work presents the use of vibration enhancement petals in order to widen the bandwidth of the energy harvester and provide a non-linear frequency response. The use of these vibration enhancement petals has allowed the fabrication of three design variations, each using an area of 1700 μm by 1700 μm. These designs have an operating bandwidth between 3.9 kHz and 14.5 kHz and can be scaled to achieve other targeted resonant frequencies.

Keywords: MEMS; energy harvesters; piezo-electricity; vibrations; aluminum nitride; process variation; non-linear; multi-mode

1. Introduction

Energy recovery systems and energy harvesting systems are the subjects of growing interest [1–5]. In addition, the number of connected objects is constantly increasing [6], and the energy requirements of these systems are still growing and many sensors are manufactured without it being physically possible or economically viable to replace their batteries once they are depleted [5]. The development of energy recovery and energy harvesting systems makes it possible to increase the lifespan of these objects, which is important notably in devices placed in remote locations or harsh environments.

Research and development of such energy harvesting systems has been focused on exploring the use of several different energy sources that can be converted into electrical energy. These energy sources need to be readily available in the environment. These include heat, light, magnetism, wind, and vibrations [7,8]. Several systems have been proposed to exploit one or more of these energy sources to produce electrical energy [9–11].

In this work, the use of vibrations as an energy source is considered by specifically focusing on piezoelectric energy harvesters. End applications include but are not limited to devices for biomedical, automotive, industrial or even military application [12–14]. When using vibrations as the main energy source, the device should be able to power the circuit in which it is integrated. Due to the nature of vibrations present in various environments, these devices need to have a low resonant frequency, and provide a sufficient power output while having a wide enough bandwidth to allow powering of the device over a wide range of vibration frequencies. These devices also need to be compact in order to allow their integrations into actual systems. However, achieving one of these objectives generally results in a challenge in maintaining the others. Therefore, a tradeoff has to be determined between the size of the device, its main resonant frequency, bandwidth, and maximum output power. Thus, the development of new structures for the implementation of piezoelectric vibration energy harvesters (PVEH) is necessary. To this end, this work will propose PVEH structures fabricated in silicon on an insulator microelectromechanical system (MEMS) fabrication process.

The most basic structure used for the realization of a MEMS PVEH is the cantilever [2,15–18]. The performance of such harvesters has been studied for energy harvesting purposes, and many different analytical models have been developed to explain and predict the behavior of this type of device. While these structures are relatively easy to fabricate and their behavior are well predicted, they are still the subject of research to overcome their limitations. Namely, the scaling down of the size of these devices proves to be a complex challenge, as their resonant frequency increases linearly with the scaling factor [19]. In addition, the smaller the device, the lower the output power.

In previous work, it has been explained how tuning the dimensions of a cantilever to attain a T-shape will allow for the reduction of up to 30% of the resonant frequency, however, this comes at the cost of reducing the power output [20]. It also has been theorized that linking four beams with a central proof mass in a cross design could be a step toward more efficient design than the cantilevers [21].

While it is possible to scale down and even tune the resonant frequency of these cantilevers by scaling them up, changing the material used for their realization, changing their geometry, using planar masses or even adding a proof mass [20], tuning the bandwidth of these designs is more difficult. While modifying the ambient pressure at which the device operates can provide such tuning, it is not practical and costly to package. The bandwidth of an energy harvester determines both its maximum output power and the frequency range in which the device can operate. This bandwidth is linked to the quality (Q) factor, and the higher the Q factor, the higher the maximum power output but the narrower the bandwidth. On the contrary, the lower the Q factor, the lower the maximum power output but the wider the bandwidth [8,14,22,23]. A trade-off in terms of the Q factor for the devices thus has to be realized in order to allow for both a sufficient output power while maintaining a wide enough bandwidth to accommodate a wide range of vibration frequencies. To overcome this trade-off, researchers have explored the design of devices using the coupling of two or more structures each with different resonant frequencies in order to allow for a wider bandwidth without compromising the maximum output power.

As such, work in [16] presents a structure allowing a maximum power output of 20 µW, however, the proposed device is large (15 mm × 15 mm), and the resonant modes of the structure are spaced between 100 Hz and 1 kHz. Furthermore, is not possible to fabricate the device using MEMS technology, limiting the miniaturization potential of the design and its batch fabrication. To reduce the resonant frequency, the design in [17] utilizes the coupling of two cantilever beams to achieve a resonant frequency of 16 Hz with a bandwidth of about 14 Hz, and a maximum power output of 0.35 µW. To achieve such performance, this work required a relatively large device (55 mm × 18 mm). Using the same concept of using several coupled cantilevers, work in [2] presents another structure with an operating frequency between 10 Hz and 20 Hz. However, once again requiring large dimensions for the devices (90 mm × 10 mm), but achieving a maximum power output of 249.8 µW. Other strategies for wideband mechanical energy harvesting are detailed in [18]. A design proposed in [18] is a cantilever occupying an area of 30 mm × 10 mm with a frequency tuning system in the range of 210 Hz to 280 Hz. Work in [24] presents the use of a tri-legged structure to harvest random vibrations and produce up to 22 V, however once again the dimensions of the structure are relatively large (diameter of 52 mm). To add to the limitations of these devices, they cannot be integrated into compact embedded systems and batch fabricated at low cost, as none of these devices were fabricated using MEMS microfabrication technology.

More compact devices using MEMS microfabrication technology are discussed below. Structures such as the one described in [1] have been reported with five resonances and 478 Hz of bandwidth, while keeping the dimensions relatively small (12 mm × 12 mm) and reporting a maximum output power of 1.73 µW. The scaling down of this structure proves to be detrimental, as illustrated by [25]. In that work, a similar structure is proposed as a vibration sensor and scaled to 2.35 mm × 2.35 mm with resonant frequencies of 14.53 kHz, 22.80 kHz and 22.81 kHz As this device is used as a vibration sensor, no output power is presented, but the work reports a sensitivity of 46 mV/g. Works in [13] and [3] present the Four-Leaf Clover, a structure with several resonances ranging from 200 Hz to several kHz and capable of delivering according to simulations a maximum power output of 11 µW at 230 Hz with dimensions of 8 mm × 8 mm. The work in [26] presents the use of structures with two resonant frequencies close to one another in order to present a much wider bandwidth. Simulations report a maximum of 10.1 µW and 9.3 µW at frequencies of 2.49 kHz and 2.55 kHz, occupying an area of 12.8 mm × 4.8 mm. Work in [5] presents the use of coupled cantilevers in an M-shape. The use of such a design allows for two resonant frequencies at about 1.29 Hz and 1.78 kHz, while the design occupies an area of 2.9 mm × 1 mm. The power output of that work is not reported.

While at the macro scale the literature seems to use cantilevers in order to reach wider bandwidths, at the micro scale, works are exploring variations of existing shapes and new shapes to overcome the limitations of the cantilever in order to increase the performances of vibrational energy harvesters [23,27]. These devices have dimensions ranging from a few millimeters to a few centimeters. Their operating frequency ranges from few hertz to several kilohertz, and their output power varies from a few nanowatts to a few microwatts.

This paper presents a new architecture for a PEVH design named herein the squared daisy. This structure is demonstrated via the PiezoMUMPS microfabrication process by Memscap. It allows for a structure that is insensitive to process variation and can operate over a relatively wide bandwidth and over different modes.

The paper is structured as follows: The operating principle of the squared daisy PEVH along with the finite element method (FEM) simulation results are detailed in Section 2. This is followed in Section 3 by the presentation of the fabrication process used for the realization of these devices and the parameters that were selected for fabricated devices. Thereafter, in Section 4, the measurement results are presented, including the characterization of the devices' variability, resonant frequency, frequency response, and mode shape. Section 5 discusses the devices and compares their performance with other works. Finally, conclusions are presented.

2. Materials and Methods

This paper introduces the squared daisy (SD) as a new geometry for the realization of MEMS energy harvesters. This structure aims to be a solution to improve the operating frequency range of MEMS energy harvesters. The visualization of such a structure is presented in Figure 1.

2.1. Operating Principle of the Squared Daisy

The SD structure can be simplified into two separate designs: A central proof mass (in orange in Figure 1) suspended by a predefined number of supporting beams (in green in Figure 1) and cantilevers attached to that central mass (in blue in Figure 1). A cross-section is illustrated in Figure 2. The central proof mass is suspended by a predefined number of supporting beams. The fundamental resonant frequency of the structure in Figure 2 can be expressed as a function of E, the modulus of elasticity of the material, I is the area moment of inertia for that structure, a function of its shape, Rm is the radius of the proof mass, H is the height of the proof mass, the density of the material ρ and L is the length of the structure [14]. It is given by:

$$f_0 = \left(\frac{1}{2\pi}\right) \times \sqrt{\frac{48 \times E \times I}{2 \times \pi \times Rm \times \rho \times H \times L}} \tag{1}$$

As governed by (1), the greater the force, the lower the resonant frequency. The values of H and a are fixed, therefore increasing the value of Rm should be favorable to reduce the resonant frequency of the device by increasing the force applied to the device. However, increasing the value of *Rm* reduces the effective length of the device, L_{eff}. This parameter is used to represent the reduction of the length of the supporting cantilevers due to the proof mass. The value of L_{eff} can be expressed in the function of the total length of the design, *Rm* and a scaling factor *Sc*. The expression of L_{eff} is given by:

$$L_{eff} = L - (2 \times Rm) \times Sc \tag{2}$$

By combining (1) and (2), the resonant frequency of the design can be shown to be given by:

$$f_0 = \left(\frac{1}{2\pi}\right) \times \sqrt{\frac{48 \times E \times I}{(2 \times \pi \times Rm \times \rho \times H) \times (L - (2 \times Rm) \times Sc)}} \tag{3}$$

Three main factors have an influence on the resonant frequency of the device: The total length of the device, the radius of the proof mass and the length of the arms. These parameters are linked by the fact that the size of the design is equal to two times the length of the arms and two times the radius of the central mass. While this basic model gives an understanding of the influence of modifying the value of *Rm*, it should be noted that it does not take into account the shapes of the cantilevers supporting the central mass and the fact that not all sides of the proof mass are supported.

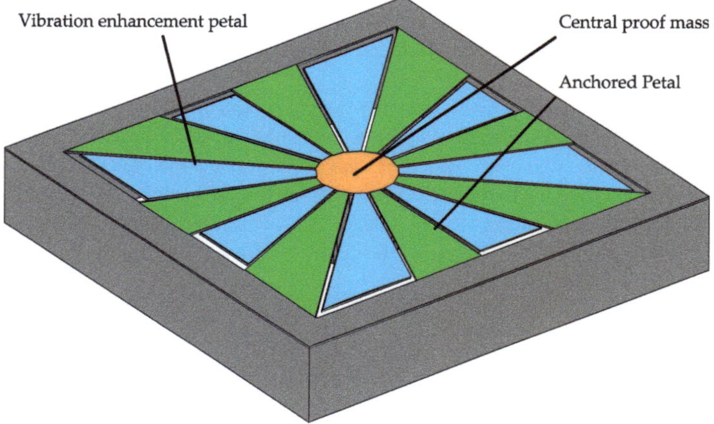

Figure 1. Visualization of the squared daisy design and identification of the main components of the daisy design: A central mass (**orange**) suspended by a predefined number of supporting beams (**green**) and the vibration enhancement petals attached to the central mass (**blue**).

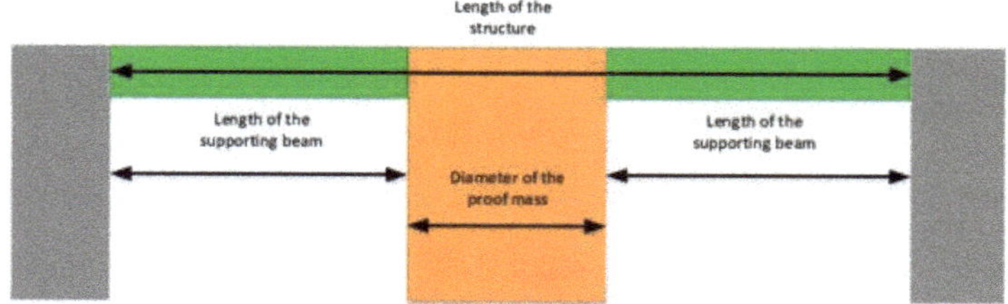

Figure 2. Visualization of a cross-section of the anchored structure.

To allow for a more complete representation of the behavior of the system, finite element modeling (FEM) of the structure has been carried out using the COMSOL Multiphysics software. The result in terms of the resonant frequency of such a system when modifying the radius of the proof mass and keeping the length of the structure constant is illustrated in Figure 3. To provide a general case, the resonant frequency has been normalized to its lowest value and the proof mass radius, *Rm*, is expressed as the percentage of the total length of the structure.

It should be noted that if *Rm* is below 5% of the length of the structure then the behavior of this structure changes and is closer to a clamped-clamped cantilever. If *Rm* is equal to 50% of the length of the structure, then the diameter of the proof mass is equal to the length of the structure, and no supporting beams remain.

Two main factors influence the resonant frequency of the system: The longer the supporting beams, the lower the frequency. However, an increase of the length of the arms also implies, for a given total length, a reduction in the radius of the proof mass and therefore an increase of the resonant frequency. It can be seen in Figure 3 that these two effects work in an opposite fashion creating a region where the influence of these two effects counteracts the other. That region is seen in Figure 3 where it can be observed that a change in radius of the proof mass between 10% to 23% of the structure length has little influence on the resonant frequency of the system. This region can be leveraged to reduce the sensitivity of the structure to process variations during device fabrication.

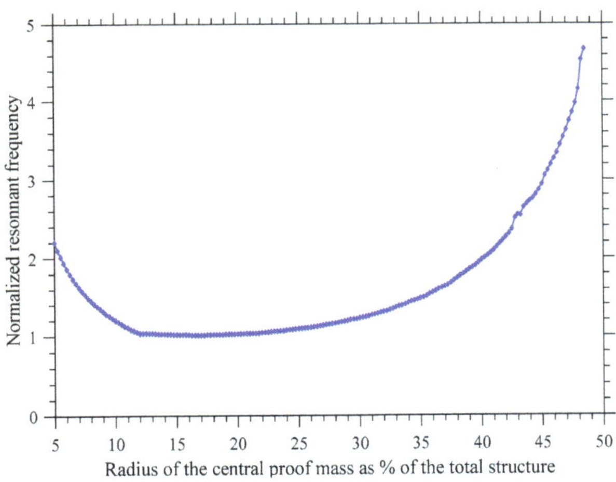

Figure 3. Simulation of the influence of the size of the proof mass on the devices.

2.2. Vibration Enhancement Petals

For the vibration enhancement petals attached to the central mass, these can be modelled as cantilevers. As such, the undamped resonant frequency in the case of a rectangular cantilever can be expressed as a function of a factor used to represent the vibration mode k_i, the length of the cantilever L, and the cross-sectional area A [8] such that:

$$f_i = \frac{k_i{}^2}{2\pi \times L^2} \times \sqrt{\frac{E \times I}{\rho \times A}} \tag{4}$$

In the case of a rectangular beam of sufficient length such that the width of the cantilever can be neglected, the fundamental resonant frequency can be expressed as a function of its length as:

$$f_i = \frac{1.875^2}{2\pi} \times \sqrt{\frac{E \times I}{\rho \times L^3}} \tag{5}$$

It should be noted that in this case, while all the vibration enhancement petals are situated around the central proof mass, they do not have the same properties, in terms of length and width at the end, it is therefore expected that these small variations will result in small resonant frequency variations that will allow for a wider overall bandwidth.

It should be noted that in a trapezoidal beam such as that of the petals, the cross-sectional area is not a constant over the length of the cantilever as the width of the cantilever is equal to *wb1* at one end and *wb2* at the other. As such, the expression of the aforementioned expressions can only give a rough estimate of the fundamental mode.

An analytical model of the SD structure is beyond the scope of this work. More advanced material on the analytical modeling of cantilevers and more complex structures can be found in [7,8,14,22,23].

These cantilevers are introduced by vibration enhancement petals (VEP), as their presence allows the addition of their resonant frequency to the global resonant behavior of the structure. Furthermore, by having the VEP supported from the central mass, the structure exhibits different properties in terms of frequency response and fundamental resonant mode, as will be discussed later.

2.3. Resilience to Process Variation

Process variation is an issue in microfabrication. In this particular case, the PiezoMUMPS process allows for an edge bias of up to 50 µm in the trenches below the silicon structure of the SD [28]. This can vary the length of the suspended structure and affect resonant behavior. Accordingly, in order to limit the influence of such process variation, two main strategies have been used. The first strategy used was to anchor the design on all sides such that the surrounding silicon structure overhangs the underlying trench at the perimeter of the structure, as demonstrated in an ultrasonic transducer demonstrated in [29]. However, while this strategy reduces the influence of the process variation on the anchoring at the perimeter of the structure, it will have no influence on process variation touching the central proof mass. This proof mass is a key component in this design to reduce the resonant frequency. Therefore, a variation of the proof mass radius by up to 50 µm would render such a design difficult to fabricate reliably. However as stated previously, in this design, a region in which the variation of the radius of the proof mass has a negligible influence on the resonant frequency of the design exists, as seen in Figure 3. This increases the resilience of the device to process variations of the SD design and ensures that the resonant frequency does not vary significantly.

3. Fabrication

In the previous section, the operating principle of the squared daisy has been overviewed. The fabrication process used to realize and validate the operation of these devices is detailed, along with the device parameters that have been selected.

The layout for the fabrication of these devices has been scripted using the SKILL language using the layout tool Cadence Virtuoso. SKILL was used to describe one parameterized cells (PCell) that contained the layout of the SD design. This PCell describes the layout of the devices, and allows for easy scaling of the device and variation of its parameters. The different layers of one variation of the SD design are shown in Figure 4, outlining each of the masks of the process. In that design all the dimensions are controlled via dimensions that are parameterized and each petal can be defined as an anchored petal or as a vibration enhancement petal.

The fabrications of these prototypes were carried-out using the PiezoMUMPs process from MEMSCAP. This process is a piezoelectric-based MEMS process that provides cost-effective access to MEMS prototyping. This process has been used and described for the creation of several resonators and energy harvesters [20,21,25,30,31]. The fabrication process includes five masks and is carried out on an N-type double-side polished silicon-on-insulator (SOI) wafer. The 10 µm-thick silicon layer (Figure 4d) is doped to increase its conductivity. Then, an insulating 0.2 µm-thick layer of silicon dioxide is grown and patterned on the SOI wafer (Figure 4c). A 0.5 µm-thick piezoelectric layer of aluminum nitride (AlN) is then deposited and patterned (Figure 4b). Then, a layer of metal is deposited, this layer consists of a stack of 0.02 µm-thick chromium (Cr) and of 1 µm-thick aluminum (Al) (Figure 4a). The silicon device layer is then patterned (Figure 4d). Lastly, the 400 µm substrate is etched from the backside to form the trench below the structure and free the proof mass to enable the structure to vibrate (Figure 4e). Further details pertaining to the fabrication process can be found in [28]. It is worth noting that this process allows for the use of the suspended substrate to be used as a proof mass. Upon reception of the device from the foundry, no post-processing step was applied to the devices.

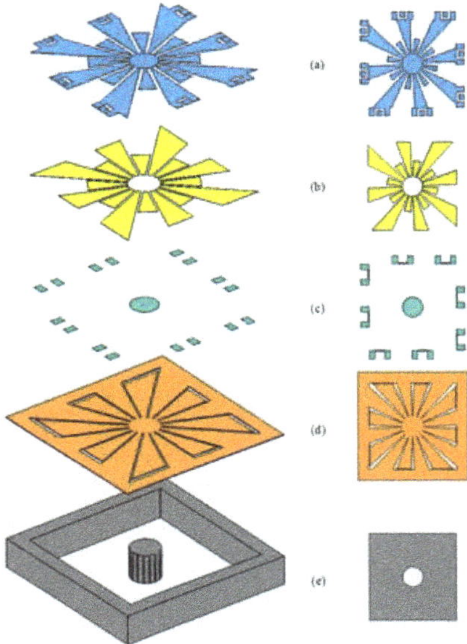

Figure 4. Illustration of the PiezoMUMPs process layers that compose the square daisy: (**a**) Metal aluminum pads, (**b**) piezoelectric AlN, (**c**) silicon dioxide insulator, (**d**) silicon device layer, and (**e**) trenched silicon substrate (left: Isometric view, right: Top view).

In order to validate the use of the SD design and the validity of the frequency enhancement petals, three different design variations were created. The parameters used to create these structures are presented in Table 1. The *size* parameter of the design represents the side of the square in which the design fits, however, this does not take into account the space needed to place the bond pads. The parameter *radius of the proof mass* represents the radius of the circle that is used to create the proof mass. The *separation on the VEP* parameter is used to determine the distance between the end of the vibration enhancement petals and the substrate. The *radius of the circle of deposited piezo* parameter is used to determine the area covered in piezo material on the VEPs. The *petals used as anchors* and *petals used as a VEP* parameter are complementary (each petal is either used as an anchor or a VEP) and are used to determine the anchoring of the structure. The use of these parameters allowed the realization of the three variants using a single PCell.

The only difference between these three variants is the way the petals are used. In the first variant, eight petals are used as anchors while eight are used as VEP. In the second and third variant, it was chosen to use only four petals as anchors while the 12 others are used as vibration enhancement petals. Pictures of the realized structures are shown in Figure 5. Fifteen copies of each of these devices were received. In that figure, the measurements points "Center", "VEP1" and "VEP2", respectfully in blue, red and orange are clearly identified for the variants and will be referred to in the measurements presented in Section 4. For the isometric view provided for variant 1, it is not possible to see the central proof mass due to the too small opening between the petals at the center of the structure.

Table 1. Overview of the parameters for each of the three variations of the SD design.

	Variant 1	Variant 2	Variant 3
Size of the design (μm)	1700	1700	1700
Radius of the proof mass, *Rm* (μm)	200	200	200
Separation on the VEP (μm)	10	10	10
Radius of the circle of deposited piezo material on VEP(μm)	500	500	500
Petals used as anchors	1 3 5 7 9 11 13 15	1 5 9 13	3 7 11 15
Petals used as a VEP	2 4 6 8 10 12 16	2 3 4 6 7 8 10 11 12 14 15 16	1 2 4 5 6 8 9 10 12 13 14 16

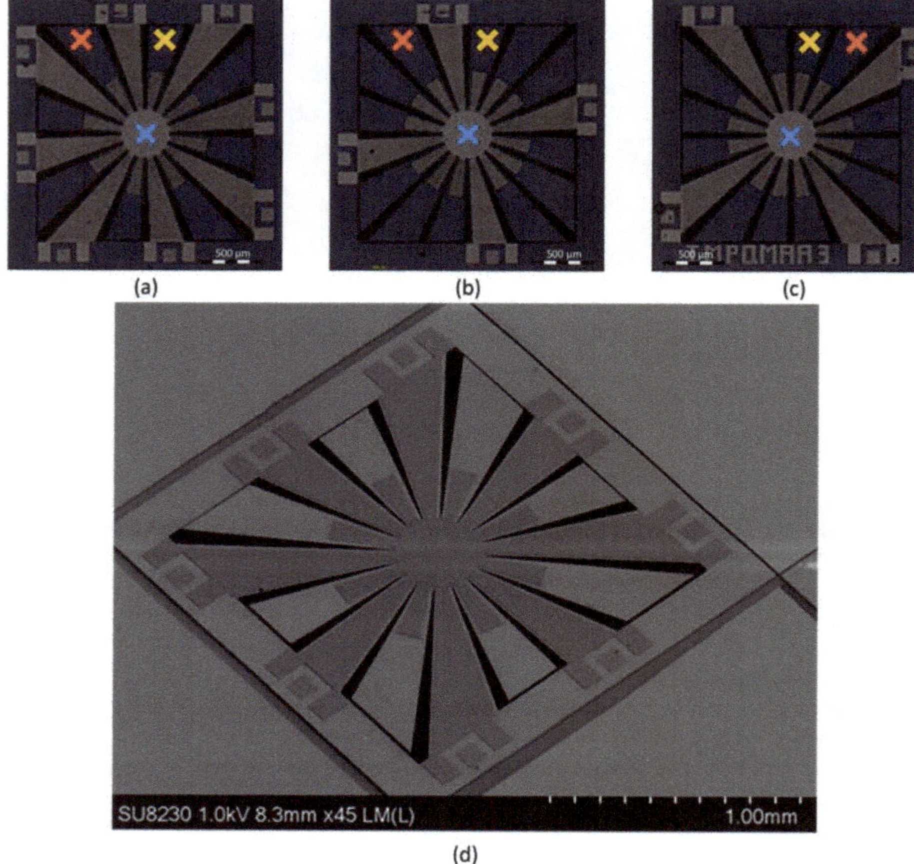

Figure 5. Presentation of the 3 fabricated designs the variant 1 in (**a**) has 8 anchored petals and 8 vibrations enhancement petals, while both variant 2 in (**b**) and variant 3 in (**c**) have 4 anchored petals and 12 vibration enhancement petals. An isometric view of variant 1 is provided in (**d**).

4. Measurement Results

In this section, the characterization of the performance of the fabricated SD designs is presented. This includes the measurement of their resonant behavior, the characterization of their frequency response in terms of power output, and the visualization of their mode shapes.

4.1. Characterization Using a Vector Network Analyzer

To characterize the resonant frequency of the designs in terms of frequency response, a vector network analyzer (VNA) type E5061B from Keysight was used with a probe station. A hole is present below the devices in order to allow the displacement of the central proof mass. By measuring the S parameters of these devices (S21 and S12), this test allowed a quick characterization of the variants. The resonant frequency of all the devices were recorded and stored, for each variant the median, average and the standard deviation of the resonant frequency was calculated. For variant 3, the extraction of the first resonant mode with the VNA was not possible due to limitation of the test setup. However, the extraction of the second mode at 7.3 kHz was possible and should be indicative of the variation of the 1st mode. The fundamental mode of variant 3 was extracted using a vibrometer, as discussed in Section 4.2. Table 2 presents a summary of these measurements. It can be noted that the good agreement of the simulations and the measurements. The maximum standard deviations in terms of resonant frequency is below 3.11% over the 15 devices measured for each variant. These results validate both of the approaches used in order to reduce the sensitivity of the structure to process variations and also validate the FEM simulations in terms of resonant frequencies.

Table 2. Summary of the resonant frequency of the three variations of the squared daisy (SD) design.

	Variant 1	Variant 2	Variant 3 *
Simulation (Hz)	10,600	8100	7300
Median (Hz)	10,680	8120	7286
Mean (Hz)	10,637	8141	7308
Standard deviation (Hz)	129	196	227
Standard deviation (%)	1.21	2.41	3.11

* Second mode.

4.2. Characterization Using a Vibrometer

A vibrometer is capable to make non-contact vibration measurements of a surface. It is possible to use it to acquire the frequency response in terms of velocity of a point. This frequency response allows the identification of the resonant mode. Furthermore, the use of a vibrometer allows the visualization of the mode shape of the device, via the measure of the speed of 2500 points at regular intervals over the area covered by the MEMS device. A data management system from Polytec is used, the vibrometer controller used is the OFV-2570 from Polytec and the laser unit is the OFV-534 from Polytec. This equipment setup is presented in Figure 6. To obtain these results, the excitation of the PVEH is provided using a function generator type 33,250A from Agilent. This excitation source is used to generate a sweep in frequency from 1 kHz to 20 kHz at a constant amplitude.

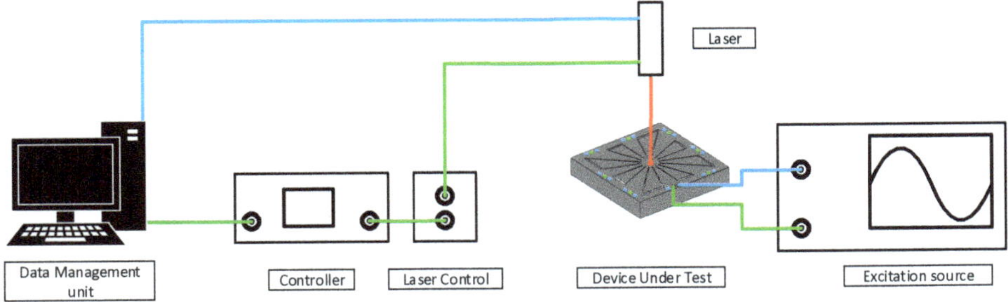

Figure 6. Schematic of the vibrometer test bench.

The vibrometer is used to acquire the displacement velocity of each point as a function of the frequency. Using a positioning controller type Corvus Evo from PI and 2 positioners type VT-80 Linear

Stage from PI. Using this equipment, 2500 points linearly spaced in a 50 × 50 grid around the design can be defined. Using this grid, the movement of the structure can be reconstructed at a defined frequency, as the exact coordinates where the points are measured are known. An example of the representation of the mode shape for variant 1 at its first resonant frequency is shown in Figure 7.

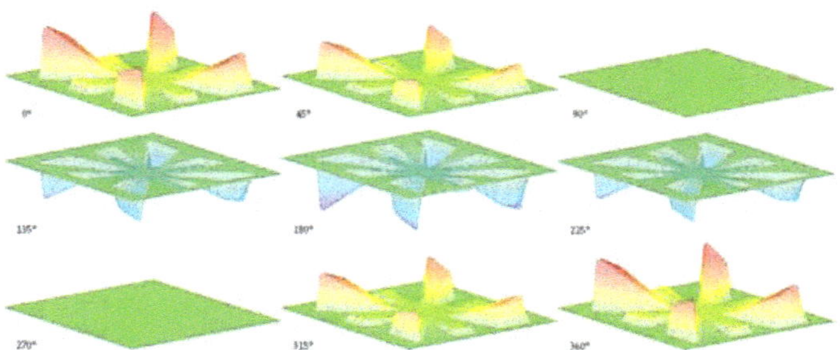

Figure 7. Representation of the 1st mode shape of variant 1 as a function of the phase.

To realize this measurement, variant 1 was exposed to a sinusoidal signal of amplitude 10 V at a frequency of 10.86 kHz. At this frequency, the resonant mode of the variant 1 is a combination of three vibration patterns. The first is caused by petals number 2, 6, 10 and 14 and presents the maximal displacement. The second is caused by petals number 4, 8, 12 and 16, with an amplitude that is lower than the first one. These two resonant patterns are different as the dimensions of these petals are different as shown in Figure 5, leading to different resonant frequencies for both. Finally, the third vibration pattern is one of the central proof mass with an amplitude that is the lowest but that also causes the displacement of the petals number 1, 3, 5, 7, 9, 11, 13 and 15.

This analysis is consistent with the data presented in Figure 8, which shows the velocity frequency response of three different locations on the device. For each of the variants, the frequency response in terms of velocity from 1 kHz to 22 kHz has been characterized at the Center, VEP1 and VEP2 points on the device, while it is subjected to a sinusoidal electrical excitation signal of 20 V amplitude. The positions of the measured points on the device are outlined in Figure 5. The test setup used is the same as described previously. These results are presented in Figure 8. It can be observed that each design possesses several resonant modes between 1 kHz and 22 kHz, in addition, several of these modes present non-linear properties. Concerning the amplitude of these modes, it is noticeable that the amplitude of the velocity of points VEP1 and VEP2 is usually higher than the one of the center. Furthermore, some modes present a higher velocity. These modes are expected to achieve a higher power output. The characteristics of the resonant modes of the central proof mass when exposed to a sinusoidal excitation signal of 20 V amplitude are summarized in Table 3. Some modes have been omitted from this table as they are not present in all of the three measured points or because they only are relatively weak, becoming apparent after a higher level of excitation has been applied. FEM simulation results of the resonant frequency have been included to highlight the good agreement between measurement and simulation.

It should be noted that for the nonlinear modes, the start frequency, the resonant frequency, and stop frequency will vary as a function of the amplitude of the excitation [32], furthermore this non-linear behavior will only occur after a sufficient level of excitation has been reached, as can be observed in Figure 9 for the first resonant mode of variant 1. In this figure, the displacement velocity of the central point is represented from 8 kHz to 13 kHz for several levels of excitation. If the amplitude of the sinusoidal excitation signal is below 5 V, then the behavior of the device is linear. However, if the amplitude is higher than 10 V then the behavior will start to appear as non-linear. This behavior is consistent with the behavior of duffing oscillators such as described in [33].

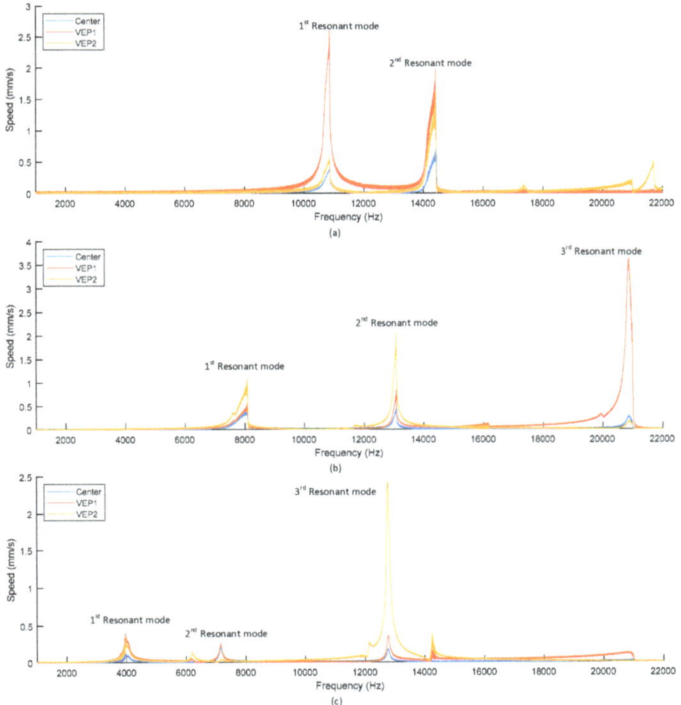

Figure 8. Velocity response of the variant 1 in (**a**), variant 2 in (**b**) and variant 3 in (**c**) from 1 kHz to 22 kHz.

Table 3. Summary of the characteristics of the resonances for the three variants at 20 V excitation.

		Variant 1	Variant 2	Variant 3
Resonant Mode 1	Start Frequency (Hz)	10,696	7466	3906
	Measured | Simulated Resonant Frequency (Hz)	10,860 | 10,800	7902 | 8100	3971 | 4050
	Stop Frequency (Hz)	10,887	7922	4138
	Linearity	Non-linear	Non-linear	Non-linear
Resonant Mode 2	Start Frequency (Hz)	14,216	12,839	7098
	Measured | Simulated Resonant Frequency (Hz)	14,485 | 15,000	12,941 | 13,100	7166 | 7243
	Stop Frequency (Hz)	14,507	12,973	7243
	Linearity	Non-linear	Non-linear	Linear
Resonant Mode 3	Start Frequency (Hz)	–	20,744	12,688
	Measured | Simulated Resonant Frequency (Hz)	–	20,871 | 20,987	12,757 | 12,833
	Stop Frequency (Hz)	–	20,987	12,833
	Linearity	–	Non-linear	Linear

These measurements, combined with the frequency response of the device validates the hypothesis about the use of several vibration enhancement petals. The petals used as VEP allow increasing the amplitude of the frequency response of the device in addition to offering additional resonant modes.

Furthermore, these petals present resonances at the same frequency as the central proof mass, this behavior should allow an improvement of the power output of the device at these frequencies.

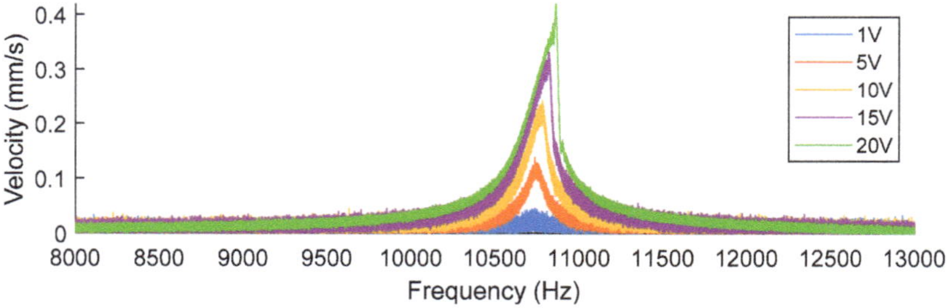

Figure 9. Frequency response around first resonant mode of variant 1 in terms of the velocity of the center point as a function of the amplitude of the excitation signal provided.

4.3. Characterization Using a Shaker

The performance in terms of output power and voltage output was characterized with a shaker type 4809 from Brüel&Kjær. This shaker is controlled by a Comet unit from Brüel&Kjær, a power amplifier type 2718 Brüel&Kjær, used with an accelerometer type 4517 from Brüel&Kjær in order to allow for proper control of the excitation signal in terms of acceleration and frequency. Regarding the data acquisition, an oscilloscope type DSO-X 3034A from Keysight is used. Once recorded, the data is submitted to an FFT algorithm to extract the amplitude of the harvested voltage. An illustration of the test setup is provided in Figure 10. However, due to frequency range operation limitations, the current test setup does not allow to characterize the performance of variant 1.

The frequency response in open circuit of variants 2 and 3, when subjected to an acceleration of 10 g, has been characterized around resonant mode 1, and is presented in Figures 11a and 12a, respectively. This data allowed the confirmation of the resonant frequency of 7.85 kHz for variant 2 where a maximum of 106 mV (amplitude) is harvested, and 3.9 kHz for variant 3 where a maximum of 64 mV (amplitude) is harvested. These frequencies will be the one used to further characterize the performances of these harvesters.

In Figures 11b and 12b, the variation of the open circuit voltage output is presented as a function of the acceleration. The apparition of non-linearity in the voltage output can be noted if the device is subjected to an acceleration greater than 6 g and 3 g, respectively for variant 2 and 3. This behavior is consistent with the data presented in Figure 9 for variant 1, and of the behavior of duffing oscillators [33]. It should be noted that accelerations of 6 g at 7.85 kHz or of 3 g at 3.9 kHz correspond in both variants to a displacement speed of about 1.2 mm/s. However, this corresponds to a displacement of about 0.15 µm in variant 2, while it corresponds to a displacement of about 0.30 µm in variant 3.

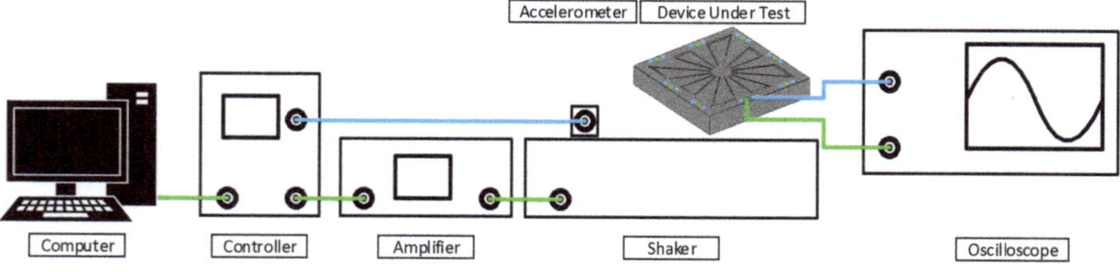

Figure 10. Presentation of the mechanical test bench for MEMS devices.

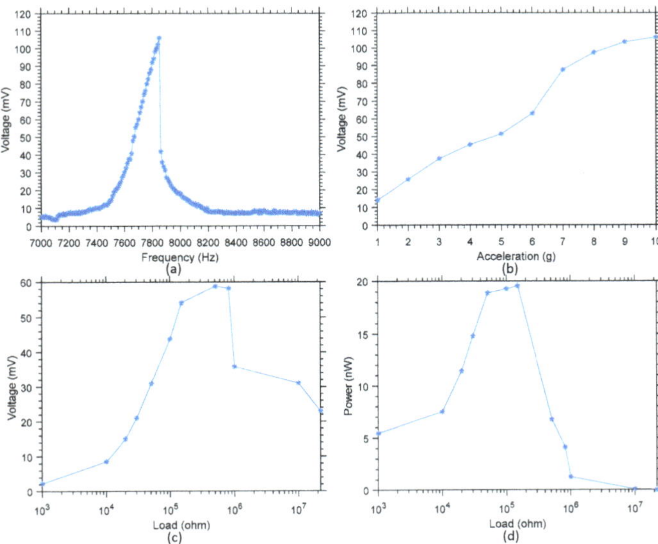

Figure 11. (**a**) Frequency response of variant 2 when subjected to an acceleration of 10 g between 7 and 9 kHz, (**b**) variation of the voltage output amplitude as a function of the input acceleration, (**c**) variation of the voltage output as a function of the load at resonance, and (**d**) and variation of the power output as a function of the load at resonance.

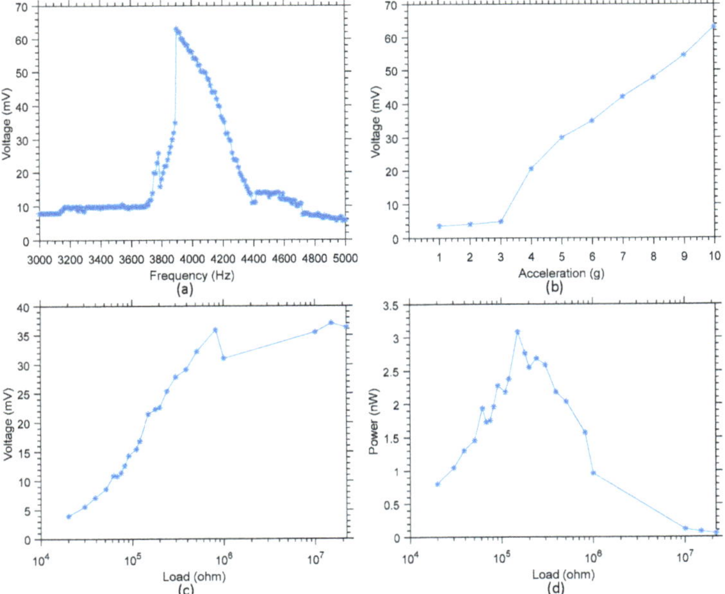

Figure 12. (**a**) Frequency response of D3 when subjected to an acceleration of 10 g between 3 kHz and 5 kHz, (**b**) variation of the voltage output amplitude as a function of the input acceleration, (**c**) variation of the voltage output as a function of the load at resonance, and (**d**) variation of the power output as a function of the load at resonance.

Finally, Figure 11c,d present the output voltage and power as a function of the load for variant 2, while Figure 12c,d present the output voltage and power as a function of the load for variant 3. For variant 2, a maximum of 20 nW is harvested, while for variant 3 a maximum of 3.1 nW of power is harvested when excited at their respective resonant frequency. It is worth noting that both these maximum were reached with a load of 150 kΩ for both devices. The difference in terms of maximum output power between these two variants is consistent with the velocity of the displacement measured and presented in Figure 8. Indeed, when excited at resonance with a similar signal amplitude, the velocity of variant 2 is much higher than that of variant 3. A higher velocity results in more displacement over a given time, and in turn more strain variation over the piezo material, resulting in higher output power.

For each of these structures, their multimode ability has been demonstrated, a visualization of the main resonant modes of the structures has been provided. In addition, their frequency response in terms of output voltage and power has been quantified. Furthermore, if a sufficient level of excitation is reached, then non-linear characteristics can be observed. This observation has been confirmed for both types of excitation: Electrical or mechanical. The low sensitivity to process the variation in terms of variation of the resonant frequency of the SD has been characterized for each variant. Concerning the reliability of the devices, these tests have occurred over a period of one year, and no device has failed during excitations that were applied. Due to their small size and high resonant frequencies, it is possible to expose them to high accelerations (~10 g) without risking their destruction.

5. Discussion

This section presents a comparison of the proposed devices with devices found in the literature. The performance of the SD design and of others are summarized and compared.

As stated in the introduction, several other designs have been proposed to overcome the limitation in terms of the operational bandwidth of cantilevers [1,13,34–36]. The performance and main characteristics of these works are summarized in Table 4. This table allows a quick overview of the performance of several devices intended to be used as MEMS PVEH energy harvesters.

Table 4. Performances of the designs and comparison with the state of the art.

	Variant 2	Variant 3	[1]	[13]	[34]	[35]	[36]
Volume (mm^3)	1.156	1.156	57.6	25.6	14.4	112	6500
Piezo material	AlN	AlN	AlN	AlN	AlN	AlN	PZT
Operating bandwidth (Hz)	7466–7922	3906–4138	730–1000	190	85–95	859.9–924.5	387–398
Linearity	Non-linear	Non-linear	Non-linear	Linear	Non-linear	Non-linear	Linear
Acceleration (g)	10	10	1	15	0,2	2	0.05
Output Voltage (mV)	54.2	21.52	770.3	130	1000	4433	1000
Load	150 kΩ	150 kΩ	70 kΩ	1 MΩ	1 MΩ	100 kΩ	1 MΩ
Output Power (nW)	20	3.1	1730	169	136	82240	52900
FOM nW·mm^{-3}·g^{-2}	0.17	0.3	30.03	0.03	236.11	183.82	3255

In order to compare the performances of these devices, one of the most common figures of merit (FOM) has been considered. The FOM expresses the maximum power output of the device in function of the volume used by the harvester and the intensity of the acceleration the device was subjected to, it is therefore expressed in nW·mm^{-3}·g^{-2}. However, it should be noted that this figure of merit has several limitations, as it is unable to express the difference between an acceleration of 10 g at 100 Hz versus an acceleration of 10 g at 10 kHz, as the first one corresponds to a displacement of 156 µm while the second one corresponds to a displacement of 0.156 µm. This difference is important as in the case of piezoelectric energy harvester, the energy is created by straining of the piezo material, and therefore the displacement magnitude has an impact on the output power. Furthermore, this figure of merit does not consider the operating bandwidth of the devices, which is an important metric of PVEH devices.

Accordingly, the FOM performance of the presented devices falls behind, as they exhibit high resonant frequencies and thus smaller displacement for a given acceleration. The performance of variant 2 is better according to this FOM than variant 3 while the operating frequency of variant 2 is about two times higher than the one of variant 3. Small devices with non-linearity and multimode capability [34] or bigger devices [36] are able to respond to very low accelerations, at low frequency, resulting in a large displacement, increased output power, and better FOM. Future work will be based on increasing the surface of piezo material where strain is present to maximize the power output of the SD device. Additional effort will be made to augment the number of resonant modes of the structures and their non-linearity. Scaling this device up in order to reduce the resonant frequency, reduces the acceleration needed to achieve sufficient displacement, and increasing the maximum power output will allow the devices to better compare with other works using the FOM.

The current dimensions, operating frequency range, response to very small displacement, and their ability to exploit several nonlinear resonances yielding operational bandwidth versatility, renders these prototyped harvester suitable for use in acoustic energy harvesting applications.

Moreover, a scale up the devices should allow the placement of proof masses on the vibration enhancement petals, in order to reduce their resonant frequency further and add more resonances to the system such as the ones observed in [3].

According to FEM simulations, if the size of the design is set to 4200 μm to occupy a full area of a block of the PiezoMUMPS process, the resonance frequency can be reduced. The volume is thus increased to 7.056 mm^3, and the operating frequency of variant 2 becomes 1.1 kHz, and that of variant 3 becomes 550 Hz. At these frequencies, the acceleration needed to obtain a similar displacement to the one presented in this table is of 0.2 g for variants 2 and 3. These values are consistent with the ones presented in [13,32,34], and indicate the potential of the design towards significantly increasing its FOM, when scaled to operate at frequencies similar to those of other works. In addition, note that this scale up does not assume that additional proof masses that could be added to a larger structure further reducing the resonant frequency.

6. Conclusions

This paper presented a new proof of concept architecture for the realization of MEMS PVEH: The squared daisy. This architecture allows for the operations of a nonlinear multi-mode energy harvester. In addition, this harvester is relatively immune to process variations due to its unique architecture. The concept of vibration enhancement petals was introduced and three variants of the designs were fabricated. The performance in terms of frequency response and variation of the power output as a function of the acceleration amplitude was given, and the visualization of the mode shape of the device, when excited at one of its resonant frequency, was provided. Future work will investigate scaling up of these devices in order to improve their power output and reduce their frequency range to allow for their use in energy harvesting applications.

Author Contributions: M.G. designed the devices, and performed the experimental testing. M.G. and A.H.A designed the experimental test setup. A.H.A. and F.N. supervised the work and provided expertise. All authors contributed to the writing of the paper.

Funding: The authors would like to thank the Natural Sciences and Engineering Research Council of Canada, Fonds de Recherche du Québec-Nature et Technologies, and the Microsystems Strategic Alliance of Quebec for their financial support.

Acknowledgments: The authors would like to thank CMC Microsystems for providing the layout design tools and enabling chip fabrication.

Conflicts of Interest: The authors declare no conflict of interest.

References

1. Jia, Y.; Du, S.; Seshia, A.A. Twenty-Eight Orders of Parametric Resonance in a Microelectromechanical Device for Multi-band Vibration Energy Harvesting. *Sci. Rep.* **2016**, *6*, 30167. [CrossRef] [PubMed]

2.   Toyabur, R.M.; Salauddin, M.; Park, J.Y. Design and experiment of piezoelectric multimodal energy harvester for low frequency vibration. *Ceram. Int.* **2017**, *43*, 675–681. [CrossRef]

3.   Iannacci, J.; Serra, E.; Sordo, G.; Bonaldi, M.; Borrielli, A.; Schmid, U.; Bittner, A.; Schneider, M.; Kuenzig, T.; Schrag, G.; et al. MEMS-based multi-modal vibration energy harvesters for ultra-low power autonomous remote and distributed sensing. *Microsyst. Technol.* **2018**, *24*, 5027–5036. [CrossRef]

4.   Nabavi, S.; Zhang, L. Frequency Tuning and Efficiency Improvement of Piezoelectric MEMS Vibration Energy Harvesters. *J. Microelectromech. Syst.* **2019**, *28*, 77–87. [CrossRef]

5.   Vyas, A.; Staaf, H.; Rusu, C.; Ebefors, T.; Liljeholm, J.; Smith, A.D.; Lundgren, P.; Enoksson, P. A Micromachined Coupled-Cantilever for Piezoelectric Energy Harvesters. *Micromachines* **2018**, *9*, 252. [CrossRef] [PubMed]

6.   Brini, O.; Deslandes, D.; Nabki, F. A System-Level Methodology for the Design of Reliable Low-Power Wireless Sensor Networks. *Sensors* **2019**, *19*, 1800. [CrossRef] [PubMed]

7.   Erturk, A.; Inman, D.J. Front Matter. In *Piezoelectric Energy Harvesting*; John Wiley & Sons, Ltd.: Hoboken, NJ, USA, 2011.

8.   Priya, S.; Inman, D.J. (Eds.) *Energy Harvesting Technologies*; Springer: Raleigh, NC, USA, 2009.

9.   Podder, P.; Constantinou, P.; Mallick, D.; Amann, A.; Roy, S. Magnetic Tuning of Nonlinear MEMS Electromagnetic Vibration Energy Harvester. *J. Microelectromech. Syst.* **2017**, *26*, 539–549. [CrossRef]

10.   Markowski, P. Thermoelectric energy harvester fabricated in thick-film/LTCC technology. *Microelectron. Int.* **2014**, *31*, 176–185. [CrossRef]

11.   Fowler, A.G.; Moheimani, S.O.R. A 4-DOF MEMS Energy Harvester Using Ultrasonic Excitation. *IEEE Sens. J.* **2016**, *16*, 7774–7783. [CrossRef]

12.   Beeby, S.P.; Tudor, M.J.; White, N.M. Energy harvesting vibration sources for microsystems applications. *Meas. Sci. Technol.* **2006**, *17*, 175–195. [CrossRef]

13.   Iannacci, J.; Sordo, G.; Serra, E.; Schmid, U. A novel MEMS-based piezoelectric multi-modal vibration energy harvester concept to power autonomous remote sensing nodes for Internet of Things (IoT) applications. In Proceedings of the 2015 IEEE SENSORS, Busan, Korea, 1–4 November 2015; pp. 1–4.

14.   Brand, O.; Dufour, I.; Heinrich, S.; Josse, F.; Fedder, G.K.; Hierold, C.; Korvink, J.G.; Tabata, O. (Eds.) *Resonant MEMS: Fundamentals, Implementation, and Application*; Wiley: Hoboken, NJ, USA, 2015.

15.   Yu, H.; Zhou, J.; Deng, L.; Wen, Z. A vibration-based MEMS piezoelectric energy harvester and power conditioning circuit. *Sensors* **2014**, *14*, 3323–3341. [CrossRef] [PubMed]

16.   Chew, Z.; Li, L. Design and characterisation of a piezoelectric scavenging device with multiple resonant frequencies. *Sens. Actuators A Phys.* **2010**, *162*, 82–92. [CrossRef]

17.   Li, S.; Crovetto, A.; Peng, Z.; Zhang, A.; Hansen, O.; Wang, M.; Li, X.; Wang, F. Bi-resonant structure with piezoelectric PVDF films for energy harvesting from random vibration sources at low frequency. *Sens. Actuators A Phys.* **2016**, *247*, 547–554. [CrossRef]

18.   Seddik, B.A.; Despesse, G.; Boisseau, S.; Defay, E. *Strategies for Wideband Mechanical Energy Harvester*; IntechOpen: London, UK, 2012.

19.   Zhu, D.; Beeby, S.P. Scaling effects for piezoelectric energy harvesters. In Proceedings of the Smart Sensors, Actuators, and MEMS VII; and Cyber Physical Systems, Barcelona, Spain, 21 May 2015; p. 95170.

20.   Alameh, A.; Gratuze, M.; Elsayed, M.; Nabki, F. Effects of Proof Mass Geometry on Piezoelectric Vibration Energy Harvesters. *Sensors* **2018**, *18*, 1584. [CrossRef] [PubMed]

21.   Alameh, A.H.; Gratuze, M.; Robichaud, A.; Nabki, F. Study and Design of MEMS Cross-Shaped Piezoelectric Vibration Energy Harvesters. In Proceedings of the 2018 25th IEEE International Conference on Electronics, Circuits and Systems (ICECS), Bordeaux, France, 9–12 December 2018; pp. 305–308.

22.   Dompierre, A.; Vengallatore, S.; Fréchette, L.G.; Vengallatore, S.; Fréchette, L.G. Piezoelectric Vibration Energy Harvesters Modeling, Design, Limits, and Benchmarking. In *Energy Harvesting with Functional Materials and Microsystems*; CRC Press: Boca Raton, FL, USA, 2013.

23.   Beeby, S.P. Energy Harvesting Devices. In *Resonant MEMS*; John Wiley & Sons, Ltd.: Hoboken, NJ, USA, 2015; pp. 451–474.

24.   Dhote, S.; Yang, Z.; Zu, J. Modeling and experimental parametric study of a tri-leg compliant orthoplanar spring based multi-mode piezoelectric energy harvester. *Mech. Syst. Signal Process.* **2018**, *98*, 268–280. [CrossRef]

25.   Yaghootkar, B.; Azimi, S.; Bahreyni, B. Wideband piezoelectric mems vibration sensor. In Proceedings of the 2016 IEEE SENSORS, Orlando, FL, USA, 30 October–3 November 2016; pp. 1–3.

26. Hoffmann, D.; Bechtold, T.; Hohlfeld, D. Design optimization of MEMS piezoelectric energy harvester. In Proceedings of the 2016 17th International Conference on Thermal, Mechanical and Multi-Physics Simulation and Experiments in Microelectronics and Microsystems (EuroSimE), Montpellier, France, 18–20 April 2016; pp. 1–6.

27. Zhu, D.; Tudor, M.J.; Beeby, S.P. Strategies for increasing the operating frequency range of vibration energy harvesters: A review. *Meas. Sci. Technol.* **2009**, *21*, 22001. [CrossRef]

28. Cowen, A.; Hames, G.; Glukh, K.; Hardy, B. *PiezoMUMPs Design Handbook*; MEMSCAP Inc.: Crolles, France, 2014.

29. Robichaud, A.; Deslandes, D.; Cicek, P.; Nabki, F. A Novel Topology for Process Variation-Tolerant Piezoelectric Micromachined Ultrasonic Transducers. *J. Microelectromech. Syst.* **2018**, *27*, 1204–1212. [CrossRef]

30. Pons-Nin, J.; Gorreta, S.; Dominguez, M.; Blokhina, E.; O'Connell, D.; Feely, O. Design and test of resonators using PiezoMUMPS technology. In Proceedings of the 2014 Symposium on Design, Test, Integration and Packaging of MEMS/MOEMS (DTIP), Cannes, France, 1–4 April 2014; pp. 1–6.

31. Elsayed, M.Y.; Nabki, F. Piezoelectric Bulk Mode Disk Resonator Post-Processed for Enhanced Quality Factor Performance. *J. Microelectromech. Syst.* **2017**, *26*, 75–83. [CrossRef]

32. Nabavi, S.; Zhang, L. Nonlinear Multi-mode Wideband Piezoelectric MEMS Vibration Energy Harvester. *IEEE Sens. J.* **2019**, *19*, 4837–4848. [CrossRef]

33. Kovacic, I.; Brennan, M.J. *The Duffing Equation: Nonlinear Oscillators and Their Behaviour*; Wiley: Hoboken, NJ, USA, 2011.

34. Rezaeisaray, M.; Gowini, M.E.; Sameoto, D.; Raboud, D.; Moussa, W. Low frequency piezoelectric energy harvesting at multi vibration mode shapes. *Sens. Actuators A Phys.* **2015**, *228*, 104–111. [CrossRef]

35. Wang, N.; Sun, C.; Siow, L.Y.; Ji, H.; Chang, P.; Zhang, Q.; Gu, Y. AlN wideband energy harvesters with wafer-level vacuum packaging utilizing three-wafer bonding. In Proceedings of the 2017 IEEE 30th International Conference on Micro Electro Mechanical Systems (MEMS), Las Vegas, NV, USA, 22–26 January 2017; pp. 841–844.

36. Aktakka, E.E.; Najafi, K. Three-axis piezoelectric vibration energy harvester. In Proceedings of the 2015 28th IEEE International Conference on Micro Electro Mechanical Systems (MEMS), Estoril, Portugal, 18–22 January 2015; pp. 1141–1144.

MDPI
St. Alban-Anlage 66
4052 Basel
Switzerland
Tel. +41 61 683 77 34
Fax +41 61 302 89 18
www.mdpi.com

Sensors Editorial Office
E-mail: sensors@mdpi.com
www.mdpi.com/journal/sensors